RELEVANT PERSPECTIVES IN GLOBAL ENVIRONMENTAL CHANGE

Edited by **Julius Ibukun Agboola**

Relevant Perspectives in Global Environmental Change
Edited by Julius Ibukun Agboola

Published by InTech
Janeza Trdine 9, 51000 Rijeka, Croatia

Publishing Process Manager Anja Filipovic

Technical Editor Teodora Smiljanic

Cover Designer InTech Design Team

Additional hard copies can be obtained from orders@intechopen.com

Relevant Perspectives in Global Environmental Change,
Edited by Julius Ibukun Agboola
p. cm.
ISBN 978-953-307-709-3

Meet the editor

Dr. Julius Ibukun Agboola is a Marine Environmental Scientist, an Ecosystem and Resource Management Specialist and an experienced University Lecturer. He recently concluded a Postdoctoral Fellowship programme with the United Nations University Institute of Advanced Studies (UNU IAS), having worked and conducted research with the UNU IAS Operating Unit Ishikawa/Kanazawa on Marine Governance and Coastal Management and Satoumi related studies. He currently lectures and carries out research at the Department of Fisheries and the Centre for Environment and Science Education (CESE), Lagos State University, Lagos, Nigeria. Dr. Agboola's research interests cover several aspects of global environmental change, including biogeochemical cycle of carbon and nitrogen in coastal ecosystems, land-ocean interaction, primary productivity, biodiversity, and examining the gradients and responses of physical, chemical and biological components of the aquatic ecosystems to environmental change. He is also interested in the application of GIS in coastal ecosystem dynamic studies, integrating social and biophysical issues for sustainability and climate change research. He holds an MSc in Environmental Resource Management and a PhD in Environmental Science from Hokkaido University, Japan.

Contents

Preface XI

Part 1 Sustainable Development in a Changing Environment 1

Chapter 1 **Relevant Issues and Current Dimensions in Global Environmental Change 3**
Julius I. Agboola

Chapter 2 **Relevant Issues for Sustainable Agriculture in Sub-Saharan Africa 19**
Ademola A. Adenle and Julius I. Agboola

Part 2 Human Health and Environmental Change Impact 37

Chapter 3 **Acid Stress Survival Mechanisms of the Cariogenic Bacterium *Streptococcus mutans* 39**
Yoshihisa Yamashita and Yukie Shibata

Chapter 4 **Human Milk: An Ecologically Functional Food 55**
Adenilda Cristina Honorio-França and Eduardo Luzia França

Part 3 Environmental Monitoring 63

Chapter 5 **Factors Controlling the Incorporation of Trace Metals to Coastal Marine Sediments: Cases of Study in the Galician Rías Baixas (NW Spain) 65**
Belén Rubio, Paula Álvarez-Iglesias, Ana M. Bernabeu, Iván León, Kais J. Mohamed, Daniel Rey and Federico Vilas

Chapter 6 **Fluorescently Labeled Phospholipids – New Class of Materials for Chemical Sensors for Environmental Monitoring 89**
George R. Ivanov, Georgi Georgiev and Zdravko Lalchev

Part 4 Biodiversity and Environmental Change 113

Chapter 7 **Effect of Metal Contamination on the Genetic Diversity of *Deschampsia cespitosa* Populations from Northern Ontario: An Application of ISSR and Microsatellite Markers 115**
Sophie Gervais and Kabwe Nkongolo

Chapter 8 **Phylogenetic Analysis of Mexican Pine Species Based on Three Loci from Different Genomes (Nuclear, Mitochondrial and Chloroplast) 139**
Carlos F. Vargas-Mendoza, Nora B. Medina-Jaritz, Claudia L. Ibarra-Sánchez, Edson A. Romero-Salas, Raul Alcalde-Vázquez and Abril Rodríguez-Banderas

Preface

As a result of the rise in awareness of the adverse environmental effects human activity has had over the centuries, environmental change has now assumed a global dimension, posing significant challenges for humanity.

This book on *Relevant Perspectives in Global Environmental Change (GEC)* presents academic discourse in environmental change from a multidisciplinary perspective, ranging from sustainable development, agriculture and human health, to environmental monitoring and biodiversity. The wide variety of studies covered in this book includes, amongst others, in-depth reviews on current dimensions in global environmental change and sustainable agricultural practices in Sub-Saharan Africa, studies on evolutionary adaptation of species and large-scale patterns of biogeography, detection of changes in the molecular environment which leads to conformational change, environmental monitoring, and adaptation and survival mechanisms of human inhabited microorganism. It combines cross-cutting research studies and reviews from different perspectives, environmental and ecological niches, and suggests possible solutions to current challenges of global environmental change.

The book is comprised of eight chapters, addressing relevant issues in environmental change. Chapter one presents a general overview of relevant issues and current dimensions in global environmental change (GEC). It clarifies the pathways/indicators of GEC, the interrelatedness of natural systems and human as driver of most GEC and the consequences of GEC on natural and human systems. As a solution, it foregrounds the need for appropriate global responses, ranging from positive policy drive and actions through innovations, and a change in institutional and environmental governance frameworks where possible, while weakening some implementation barriers devastating the existing institutions. It also emphasizes the need for more interdisciplinary and integrative perspective on global environmental change issues. In a similar vein, chapter two gives an overview of relevant issues relating to sustainable agricultural development in Sub-Saharan Africa. It also examines current literature on the issues and constraints associated with practicing sustainable agriculture in this region. It further emphasizes how policy interventions through governments and international initiatives can ensure adoption of sustainable agriculture towards improving food security, agricultural productivity and environmental sustainability.

Chapter three presents a molecular study on the Acid stress survival mechanisms of the cariogenic bacterium *Streptococcus mutans,* the major etiological agent in human dental caries, from a different perspective. The authors described the acid tolerance characteristics of the cariogenic microorganism, *S. mutans* and concluded that Two-component systems (TCSs), prokaryote-specific signal transduction systems, are widespread in prokaryotes and play extensive roles in adaptation to environmental changes, and in their opinion, TCSs seem to be the most suitable system for adaptation to environmental conditions. Similarly, in chapter four, the authors describe human breast-milk as an ecologically functional food and conclude that breastfeeding is an ecological act because it supports an environment for current and future generations that enhances their quality of life and lowers the risk of malnutrition and infection.

The main factors controlling the incorporation of metals to the sediments in transitional coastal environments, like the Galician Rías Baixas in the NW Spain, is presented in chapter five. The authors emphasize the need to understand the behaviour of trace metals in sediments in order to improve coastal risks prevention and management, as well as to reach a good status in water quality as one of the great challenges for the European Union in the new millennium. In chapter six, the authors stated that reliable environmental monitoring strongly depends on the quality of chemical and biochemical sensors and propose a new class of materials – fluorescently labelled phospholipids, which can be used as chemical and biological sensors.

Furthermore, the study reported in chapter seven determines the effect of long term exposure of *Deschampsia cespitosa* populations to metal, on genetic diversity using Inter-Simple Sequence Repeat (ISSR) and microsatellite markers. The ISSR data suggested that adaptation to metal-contaminated soils in populations of *D. cespitosa* could be associated with a significant reduction in genetic variation. The study concludes that monitoring the genetic diversity of *D. cespitosa* populations has been useful in detecting trends that should alert ecologists to potential problems. Also, the high genetic variability detected with micro-satellite markers in the *Deschampsia* populations from Sudbury and Cobalt suggests that these are healthy populations with a evolutionary potential to respond favourably and adapt to changes or disturbances in the environment. Chapter eight presents a concise overview of the evolution and phylogenetic considerations in pine trees and problems concerning areas of exploitation, especially with the genus *Pinus* which are of great ecological importance as primordial members of temperate forests, and are also economically important, being a source of wood and resins. Knowing which species are more closely related and which groups show a higher diversity enables such species to be considered as those of greatest potential use, and fundamental for the integral management of forests in the face of global environmental change.

In conclusion, it presents a dynamic evolution that should intrigue the development of mechanisms to cope with present and possible future environmental change.

Julius Ibukun Agboola, Ph.D
Department of Fisheries, Faculty of Science,
Centre for Environment and Science Education (CESE),
Lagos State University, Ojo, Lagos
Nigeria

Part 1

Sustainable Development in a Changing Environment

1

Relevant Issues and Current Dimensions in Global Environmental Change

Julius I. Agboola[1,2]
[1]United Nations University, Institute of Advanced Studies, Operating Unit in Ishikawa/Kanazawa, 2-1-1 Hirosaka, Kanazawa, Ishikawa,
[2]Department of Fisheries, Faculty of Science/ Centre for Environment and Science Education (CESE), Lagos State University, Ojo, Lagos,
[1]Japan
[2]Nigeria

1. Introduction

Global environmental change (GEC) includes both systemic changes that operate globally through the major systems of the geosphere-biosphere, and cumulative changes that represent the global accumulation of localized changes. The importance and awareness of GEC has greatly increased since the second UN Conference on Environment and Development (UNCED) in 1992. During the last two decades, GEC research programs around the world have advanced our understanding of the Earth's ever-changing physical, chemical, and biological systems and the growing human influences on these systems. On the basis of current knowledge attention is now focused on the critical unanswered scientific questions that must be resolved to fully understand and usefully predict future's GEC. It is hoped that measurable significant progress would be made in the forthcoming Earth Summit 2012, formerly known as United Nations Conference on Sustainable Development (UNCSD) or Rio+20 scheduled for Rio de Janeiro in June 2012.
Generally, the earth's climate system varies naturally across a range of temporal scales, including seasonal cycles, inter-annual patterns such as the El Niño/La Niña-Southern Oscillation- ENSO, inter-decadal cycles such as the North Atlantic and Pacific Decadal oscillations, and multimillenial-scale changes such as glacial to inter-glacial transitions (Harley et al., 2006). This natural variability is reflected in the evolutionary adaptations of species and large-scale patterns of biogeography. In all, human activities play an important part in virtually all natural systems and are forces for change in the environment at local, regional, and even global scales. Human drivers of GEC include consumption of energy and natural resources, technological and economic choices, culture, and institutions. The effects of these drivers are seen in population growth and movement, changes in consumption, de- or reforestation, land-use change, and toleration or regulation of pollution, and other issues highlighted in section two of this chapter. For instance, the Intergovernmental Panel on Climate Change (IPCC) reports that, if global average temperatures exceed 2°C there will be irreversible impacts on water, ecosystems, food, coastal zones and human health. We have a 50% chance of avoiding a 2°C warming if we stabilize greenhouse gases at 450 ppm CO_2 eq (parts per million carbon dioxide equivalents). Recent evidence suggests even more rapid

change, which will greatly, and in some case irreversibly, affect not just people, but also species and ecosystems. The schematic framework on the current state of GEC, representing anthropogenic drivers, impacts of and responses to climate change, and their linkages is presented in Figure 1.

In the light of the above overview on GEC, the rest of this chapter is organized as follows: First, in section 2, a review of relevant issues in GEC is given. Section three elucidates on pathways/indicators of GEC. The fourth section centers on the interrelatedness of natural systems and human as driver of most GEC. Some of the consequences of GEC on natural and human systems are reviewed in section five. Section six dealt on global responses, ranging from positive policy drive and actions through innovations and a change where possible in institutional and environmental governance frameworks while weakening some implementation barriers ravaging existing institutions. Lastly, section seven concludes and foregrounds on the need for more interdisciplinary and integrative perspective on global environmental change issues. It hopes to proffer some definitions and answers to overarching questions in GEC.

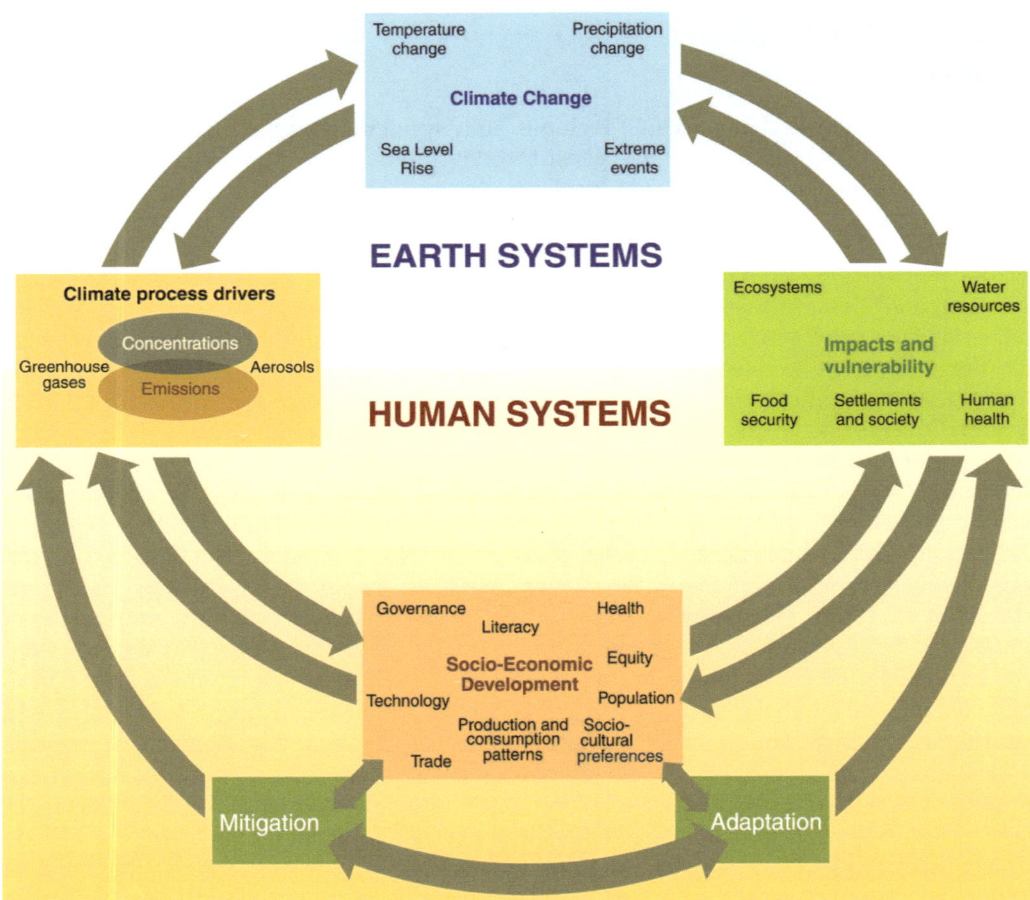

Fig. 1. Schematic framework of anthropogenic climate change drivers, impacts and response. Source: IPCC, 2007.

2. Relevant issues in global environmental change

In the past, several issues have been addressed on GEC. Very striking and concise in my opinion is Vitousek (1994). He concluded that three of the well-documented global changes are: increasing concentrations of carbon dioxide in the atmosphere; alterations in the biogeochemistry of the global nitrogen cycle; and ongoing land use/land cover change, and are perhaps the bedrock of other relevant issues in GEC mentioned in sub-sections 2.0 below. Human activity-now primarily fossil fuel combustion- has increased carbon dioxide concentrations from 280 ppm to 355 ppm since 1800; the increase is unique, at least in the past 160 000 yr, and several lines of evidence demonstrate unequivocally that it is human-caused (Vitousek, 1994; Ebi, 2011). The global nitrogen cycle has been altered by human activity to such an extent that more nitrogen is fixed annually by humanity (primarily for nitrogen fertilizer, also by legume crops and as a byproduct of fossil fuel combustion) than by all natural pathways combined. This added nitrogen alters the chemistry of the atmosphere and of aquatic ecosystems, contributes to eutrophication of the biosphere, and has substantial regional effects on biological diversity in the most affected areas. Finally, human land use/land cover change has transformed one-third to one-half of Earth's ice-free surface. This in and of itself probably represents the most important component of global change now and will for some decades to come; it has profound effects on biological diversity on land and on ecosystems downwind and downstream of affected areas. These three and other equally certain components of global environmental change are the primary causes of anticipated changes in climate, and of ongoing losses of biological diversity (Vitousek, 1994, Sala et al, 2000).

2.1 Climate change

Climate change is one of several large-scale environmental changes to which human activities make a significant contribution, and that, in turn, affects human health and well-being. The Intergovernmental Panel on Climate Change (IPCC) concluded: "warming of the climate system is unequivocal, as is now evident from observations of increases in global average air and ocean temperatures, widespread melting of snow and ice, and rising global average sea level (IPCC 2007)." Over the past decade, the fact that emissions of greenhouse gases due to human activities are affecting the world's climate has become clear (Ebi 2011). In addition: "most of the observed increase in globally averaged temperatures since the mid-twentieth century is very likely due to the observed increase in anthropogenic greenhouse gas concentrations." These changes have begun to affect morbidity and mortality worldwide, with projections suggesting that overall health burdens will increase with increasing climate change. Although all countries are projected to experience increased health risks, those at greatest risk include the urban poor, older adults, children, traditional societies, subsistence farmers, and coastal populations, particularly in low-income countries (Ebi 2011).

The task of understanding climate change and predicting future change would be complex enough if only natural forcing mechanisms were involved. It is significantly more daunting because of the introduction of anthropogenic forcing and even more so considering the limitations in available records.

2.2 Land use

Landscapes are changing worldwide, as natural land covers like forests, grasslands, and deserts are being converted to human-dominated ecosystems, including cities, agriculture, and forestry. Between 2000-2010, approximately 13 million hectares of land (an area the size of Greece) were converted each year to other land cover types (FAO 2010). Developed regions like the US and Europe experienced significant losses of forest and grassland cover over the past few centuries during phases of economic growth and expansion. More recently, developing nations have experienced similar losses over the past 60 years, with significant forest losses in biologically diverse regions like Southeast Asia, South America, and Western Africa.

Land use changes affect the biosphere in several ways. They often reduce native habitat, making it increasingly difficult for species to survive. Some land use changes, such as deforestation and agriculture, remove native vegetation and diminish carbon uptake by photosynthesis as well as hasten soil decomposition, leading to additional greenhouse gas release. Almost 20% of the global CO_2 released to the atmosphere (1.5–2 billion tons of carbon) is thought to come from deforestation.

2.3 Biodiversity

Biodiversity is the diversity of life on Earth and includes the richness (number), evenness (equity of relative abundance), and composition (types) of species, alleles, functional groups, or ecosystems. The period since the emergence of humans has displayed an ongoing biodiversity reduction and an accompanying loss of genetic diversity. Named the Holocene extinction, the reduction is caused primarily by human impacts, particularly habitat destruction.

Currently, global biodiversity is changing at an unprecedented rate and scale in response to human-induced perturbation of the Earth System. Fossil records indicate that the background extinction rate (that is Pre-Industrial value) for most species is 0.1-1 extinctions per million species per year (Braimoh et al., 2010). Over the past years however, the species extinction rate has increased to more than 100 extinctions per million species per year (MA, 2005). There is a strong linkage between biodiversity loss and human-driven ecosystem processes from local to regional scales. Conversely, biodiversity impacts human health in a number of ways, both positively and negatively (Sala et al. 2009), and there is considerable evidence that contemporary biodiversity declines will lead to subsequent declines in ecosystem functioning and ecosystem stability (Naeem et al. 2009).

Generally, observed changes in climate have already adversely affected biodiversity at the species and ecosystem level, and further changes in biodiversity are inevitable with further changes in climate (CBD 2009). While human actions have significantly contributed to the loss of biodiversity, in some cases, human actions have promoted biodiversity. Conservation strategies, such as creating parks to protect biodiversity hotspots, have been effective but insufficient (Bruner et al. 2001). For example, although biodiversity is often greater inside than outside parks, species extinctions continue. Specifically, biodiversity and ecosystem services are greater in restored than in degraded ecosystems but lower in restored than in intact remnant ecosystems (Benayas et al. 2009). Despite the positive effects of conservation and restoration efforts, biodiversity declines have not slowed (Butchart et al. 2010). Thus, further investigation is needed to determine new conservation and restoration strategies.

Global agreements such as the Convention on Biological Diversity, give "sovereign national rights over biological resources" (not property). The agreements commit countries to "conserve biodiversity", "develop resources for sustainability" and "share the benefits" resulting from their use. Sovereignty principles can rely upon what is better known as Access and Benefit Sharing Agreements (ABAs). The Convention on Biodiversity implies informed consent

between the source country and the collector, to establish which resource will be used and for what, and to settle on a fair agreement on benefit sharing. Theoretical and empirical studies have identified a vast number of natural processes that can potentially maintain biodiversity. Biodiversity can be maintained by moderately intense disturbances that reduce dominance by species that would otherwise competitively exclude subordinate species. For example, selective grazing by bison can promote plant diversity in grasslands (Collins et al. 1998).
Now, it may be possible to predict future changes in biodiversity, ecosystem functioning, and ecosystem stability by considering how global ecosystem changes are currently influencing stabilizing species interactions. In this direction, the United Nations is currently developing an Intergovernmental Science-Policy Platform on Biodiversity and Ecosystem Services (IPBES) to monitor biodiversity and ecosystem services worldwide (Marris 2010). The IPBES will be modelled after the Intergovernmental Panel on Climate Change (IPCC), and there is great potential for ecologists to borrow strategies that have been successfully employed by climatologists. Another global effort in this direction (conservation of biodiversity) recently culminated in the United Nations designating 2011-2020 as the United Nations Decade on Biodiversity during the Aichi COP 10 meeting in 2010- the International Year of Biodiversity.

3. Pathways and indicators of global environmental change

Changes are occurring throughout the Earth System and are evident in the oceans, on the land and in the atmosphere. These changes are increasingly driven by human activities. There is the need to understand how these ecosystems react to global change so as to understand the consequences for their functioning and to manage ecosystems resources sustainably. Sala et al., (2000) recognize five major drivers of biodiversity loss, namely land use, climate, nitrogen deposition, biotic exchange and atmospheric carbon dioxide, and opined that the importance of these drivers varies from one ecosystem to the other. Here, I present and elucidate the pressures on three major pathways of global environmental change.

3.1 Terrestrial

Land-use change (especially deforestation) and climate change generally have the greatest impact for terrestrial ecosystems, whereas biotic exchange is more important for freshwater ecosystems. As earlier mentioned, human activity–now primarily fossil fuel combustion–has increased carbon dioxide concentrations from 280 ppm to 355 ppm since 1800; the increase is unique, at least in the past 160 000 yr, and several lines of evidence demonstrate unequivocally that it is human-caused (Vitousek, 1994). This increase is likely to have climatic consequences–and certainly it has direct effects on biota in all Earth's terrestrial ecosystems. Land use and land cover change has aroused increasing attention of scientists worldwide since 1990. Recognizing the importance of this change to other global environmental change and sustainable development issues, the International Geosphere-Biosphere Programme (IGBP) and the Human Dimensions of Global Environmental Change Programme (IHDP) initiated a joint core project Land Use and Land Cover Change (LUCC) and published a Science/Research Plan for the project. This precipitated into a number of IGBP core projects, of which one is Global Change and Terrestrial Ecosystems (GCTE). Despite these achievements, agricultural activities have continued to be significant emitters of global greenhouse gases (GHGs) and as such agricultural activity is a major driver of

anthropogenic climate change. Emissions from agricultural sources were 14% of global GHG emissions in 2000 with developing countries accounting for three quarters of agriculture emissions in the case of rice (WRI, 2006; Stern, 2007).
Also, changes in vegetation structure influence the magnitude and spatial pattern of the carbon sink and, in combination with changing climate, also freshwater availability (runoff). The potential for terrestrial ecosystems to absorb significant amounts of CO_2, thus slowing the buildup of CO_2 in the atmosphere and reducing the rate of climate change, is a key issue in the debate on CO_2 emission controls. As more land is converted to agriculture, there is less area in natural ecosystems that can act as carbon sinks, thereby reducing the potential sink strength of the terrestrial biosphere.

3.2 Aquatic

The ocean is a vital component of the metabolism of the Earth and plays a key role in global change. In 1987 the World Commission on Environment and Development (Brundtland Commission) warned in the final report, Our Common Future, that water was being polluted and water supplies were overused in many parts of the world. The ocean is the source of most of the world's precipitation (rainfall and snowfall), but people's freshwater needs are met almost entirely by precipitation on land (see Figure 2), with a small though increasing amount by desalination. Due to changes in the state of the ocean, precipitation patterns are altering, affecting human well-being. Asides from this, ocean changes are also affecting marine living resources and other socio-economic benefits on which many communities depend, and anthropogenically induced global climate change has profound implications for marine ecosystems and the economic and social systems that depend upon them (Harley et al., 2006).
Human pressures at global to basin scales are substantially modifying the global water cycle, with some major adverse impacts on its interconnected aquatic ecosystems- freshwater and marine - and therefore on the well-being of people who depend on the services that they provide. Like other ecosystems, marine and coastal areas are already adversely impacted by many stresses, which will be exacerbated by climate change (e.g., sea level rise) (see figure 3). At the same time, coastal ecosystems ranging from Polar Regions to Small Island developing States are essential to our capacity to respond to projected climate change impacts.
In recent years, a major observed pressure on the aquatic ecosystem is Ocean acidification caused by decrease in the pH and increase in acidity of the Earth's oceans as a result of uptake of anthropogenic carbon dioxide (CO_2) from the atmosphere. Past, present and future predicted average surface ocean pH is shown in Table 1. Dissolving CO_2 in seawater increases the hydrogen ion (H^+) concentration in the ocean, and thus decreases ocean pH. Caldeira and Wickett (2003) placed the rate and magnitude of modern ocean acidification changes in the context of probable historical changes during the last 300 million years.
Furthermore, in terms of resources, aquatic ecosystems continue to be heavily degraded, putting many ecosystem services at risk, including the sustainability of food supplies and biodiversity. Global marine and freshwater fisheries show large scale declines, caused mostly by persistent overfishing. Freshwater stocks also suffer from habitat degradation and altered thermal regimes related to climate change and water impoundment. A continuing challenge for the management of water resources and aquatic ecosystems is to balance environmental and developmental needs. It requires a sustained combination of technology, legal and institutional frameworks, and, where feasible, market-based approaches.

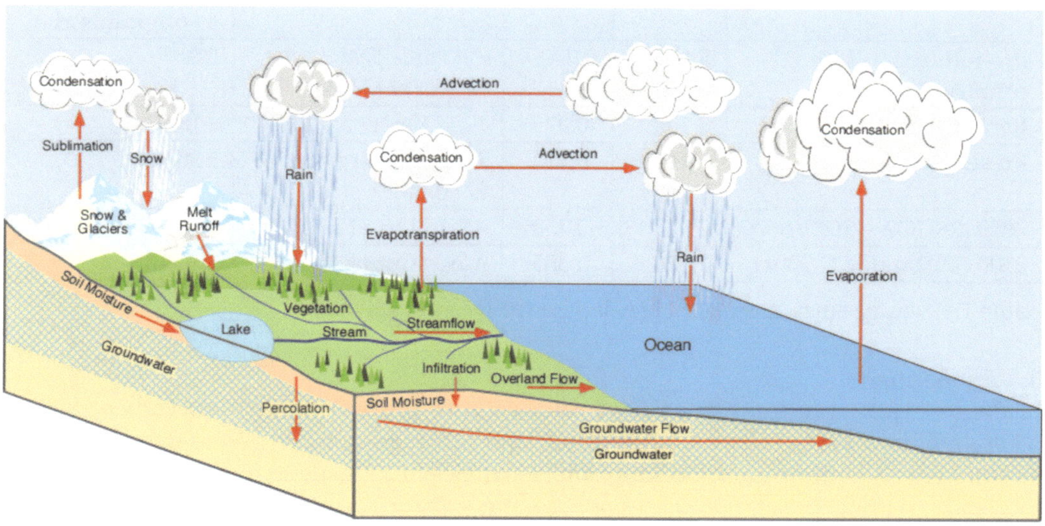

Fig. 2. Hydrologic cycle (Adapted from Pidwirny, 2006).

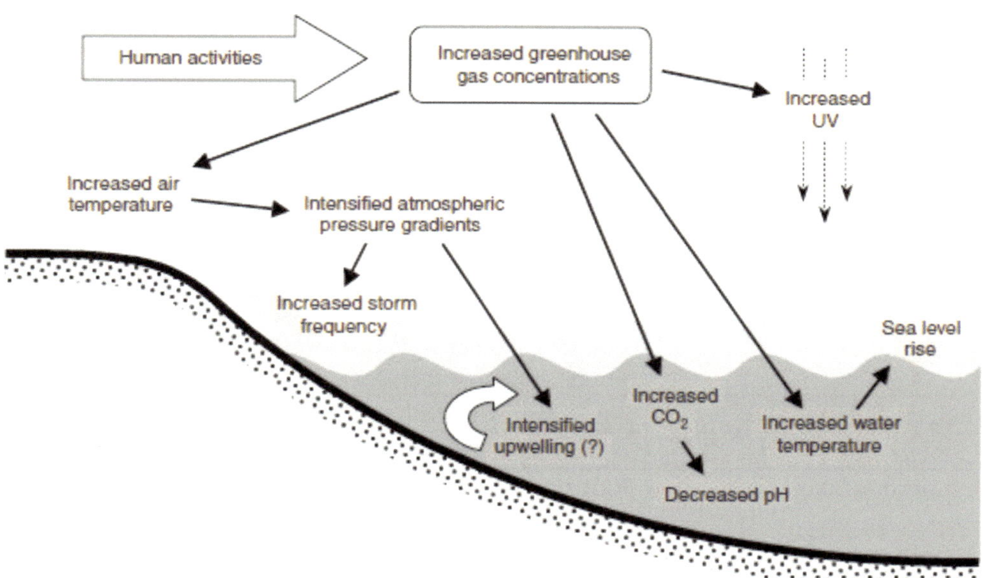

Fig. 3. Important abiotic changes associated with climate change. Human activities such as fossil fuel burning and deforestation lead to higher concentration of greenhouse gases in the atmosphere, which in turn leads to a suit of physical and chemical changes in coastal ocean. (Adapted from Harley et al., 2006).

Time	pH	pH change	Source	H^+ concentration change relative to pre-industrial
Pre-industrial (18th century)	8.179	0.000	*Analyzed field* Key et al (2004)	0%
Recent past (1990s)	8.104	-0.075	*Field* Key et al (2004)	+18.9%
Present levels	~8.069	-0.11	*Field* Hall-Spencer et al (2008)	+28.8%
2050 (2×CO2 = 560 ppm)	7.949	-0.230	*Model* Orr et al (2005)	+69.8%
2100 (IS92a) (IPCC 2001)	7.824	-0.355	*Model* Orr et al (2005)	+126.5%

Table 1. Average surface ocean pH (Adapted from Orr et al 2005).

3.3 Atmospheric

In the last two centuries, human have released ever greater quantities of carbon dioxide (CO_2), methane (CH_4), nitrous oxide (N_2O) and other greenhouse gases, into the Earth's atmosphere (Woodward and Buckingham, 2008). Scientists have found that the four most important variable greenhouse gases, whose atmospheric concentrations can be influenced by human activities, are carbon dioxide (CO_2), methane (CH_4), nitrous oxide (N_2O), and chlorofluorocarbons (CFCs). Greenhouse gases basketed under the Kyoto Protocol and their main generators are shown in Table 2. Historically, CO_2 has been the most important, but those other atmospheric trace gases are also radiatively active, in that they can affect Earth's heat budget and thereby contribute to a greenhouse warming of the lower atmosphere. CO_2 has risen from pre-industrial concentration of 280 ppm to current values in excess of 380 ppm, and is currently rising by 1.9 ppm per year (Woodward and Buckingham, 2008). There is a growing concensus among climate researchers that these greenhouse gases are causing the Earth's temperature to rise. Scientists have measured a temperature rise of 0.76 ℃ (with confidence intervals of 0.56 to 0.92 ℃) between 1850 and 2005, as a result of increased radiative forcing from the increases in atmospheric greenhouse gases (IPCC 2007).

Greenhouse Gas	Main Sources
Carbon dioxide (CO_2)	Fossil fuel combustion (e.g. road transport, energy industries, other industries, residential, commercial and public sector); forest clearing
Methane (CH_4)	Agriculture, landfill, gas leakage, coal mines
Nitrous oxide (N_2O)	Agriculture, industrial processes, road transport, others
Perfluorocarbons (PFCs)	Industry (e.g. aluminium production, semi-conductor industry)
Hydrofluorocarbons (HFCs)	Refrigeration gases, industry (as perfluorocarbons)
Sulphur hexafluoride ($SF/_6$)	Electrical transmissions and distribution systems, circuit breakers, magnesium production

Source: UNFCC, 2003

Table 2. Greenhouse gases basketed under the Kyoto Protocol and their main generators (Note: greenhouse gases produced by air transport are exempted from the Protocol).

Current developments in atmospheric chemistry are revealing the close links between chemistry, radiation, dynamics, and climate. Examples include the powerful role played by aerosol formation in both the boundary layer and the upper troposphere, chemical initiation of subvisible cirrus in the region of the tropopause, the control exerted by water vapor and temperature on the sharply nonlinear partitioning of halogen and hydrogen radicals in the lower stratosphere, and the importance of stratosphere-troposphere exchange on the composition and meteorology of the upper troposphere and lower stratosphere.

4. Human dimensions of global environmental change

According to Jager (2002), research on the human dimensions of global environmental change is concerned with the human causes of change, the consequences of such changes for individuals and societal groups, and the ways in which humans respond to the changes. The human causes include emissions of pollutants into the atmosphere, especially carbon dioxide, chlorofluorocarbons and acidifying substances, as well as land-use and land-cover changes.

It is now established wisdom that humans are the prime drivers of change on Earth, and it is this recognition that underpins the discussion of the Anthropocene. Social, economic, and cultural systems are changing in a world that is more populated, urban, and interconnected than ever. Such large-scale changes increase the resilience of some groups while increasing the vulnerability of others.

With the impetus of global change research, study of large-scale ecosystems has become a rapidly maturing field of science and has shown major successes over the past decade. Improved fundamental understanding of marine and terrestrial ecosystems and hydrology has already led to practical applications in weather and climate modeling, air quality, and better management and natural hazards responses for water, forest, fisheries, and rangeland resources. The development of spatially resolved global-scale ecosystem models has occurred only during the past five years. Computing capability and remote sensing technology have further driven change in the nature of the field. The capability has emerged not only to model at global scales but also to exploit data at these scales.

Such capability is increasingly important for developing our economy, protecting our environment, safeguarding our health, and negotiating international agreements to ensure the sustainable development of the global community of nations. As earlier mentioned, if global average temperatures exceed 2°C there will be irreversible impacts on water, ecosystems, food, coastal zones and human health (IPCC 2007). We have a 50% chance of avoiding a 2°C warming if we stabilize greenhouse gases at 450 ppm CO_2 eq (parts per million carbon dioxide equivalents). This means we must start radically reducing emissions now and stay on a low emissions pathway to avoid increasing the amount of CO_2 in the atmosphere. The good management of ecosystems such as wetlands and forests remains an effective mitigation option given the high sequestration potential of natural systems. The permanence of carbon sinks is also tied to the maintenance or enhancement of the resilience of ecosystems (CBD 2009).

5. Consequences for natural and human systems and responses

There are consequences for natural and human systems in GEC events. Studies have focused, for example, on global environmental change impacts on agriculture and human health and on particular locations, such as the coastal zone. Rising sea levels; increased temperatures; increased risk of droughts, floods and fires; stronger storms and increased storm damage; changing landscapes; forced environmental migrations and food insecurity are but a few of the issues linked to a changing climate. However, while a large proportion of climate change impacts will be negative, some will be positive too. For certain societies these will include, among others, increased agricultural growing periods and lower winter mortalities (warmer winters), although it is generally accepted that the negatives will significantly outweigh the positives (Nelson et al, 2009). The Stern Review suggests that all countries will be affected by climate change, but it is the poorest countries that will suffer earliest and most. Unabated climate change may risks raising average temperatures by over 5°C from pre-industrial levels. Such changes would transform the physical geography of our planet, as well as the human geography- how and where we live our lives (Stern Review, 2007). Some examples of these consequences are elucidated.

5.1 Food security

Food security is the ability of people to have access to sufficient, nutritious food. Global environmental change (GEC), including land degradation, loss of biodiversity, changes in hydrology, and changes in climate patterns resulting from enhanced anthropogenic emission of greenhouse gas emissions, will have serious consequences for food security, particularly for more vulnerable groups (Ericksen et al. 2009). Growing demands for food in turn affect the global environment because the food system is a source of greenhouse gas emissions and nutrient loading, and it dominates the human use of land and water. The speed, scale and consequences of human-induced environmental change are beyond previous human experience, and thus science has a renewed responsibility to support policy formation with regard to food systems (Carpenter et al., 2009; Steffen et al., 2003). Most research linking global environmental change and food security focuses solely on agriculture: either the impact of climate change on agricultural production, or the impact of agriculture on the environment, e.g. on land use, greenhouse gas emissions, pollution and/ or biodiversity (Ericksen et al. 2009).

Although, we currently grow enough food to feed the global human population, a population rising to 9 billion by 2050, combined with climate changes, will strain the capacity of some regions to feed people, thereby raising the risks of food insecurity (Godfray et al. 2010). Thus, the effects of global environmental change (GEC) are increasingly making the practical achievement of food security more difficult in some of the world's poorest communities. It is important to note that while technical fixes are important, they will not alone solve the food security challenges. Adapting to the additional threats to food security arising from major environmental changes requires an integrated food system approach, not just a focus on agricultural practices. In this line, Ericksen et al (2009) further highlighted on six key issues that has emerged for future research: (i) adapting food systems to global environmental change requires more than just technological solutions to increase agricultural yields; (ii) tradeoffs across multiple scales among food system outcomes are a pervasive feature of globalized food systems; (iii) within food systems, there are some key underexplored areas that are both sensitive to environmental change but also crucial to understanding its implications for food security and adaptation strategies; (iv) scenarios specifically designed to investigate the wider issues that underpin food security and the environmental consequences of different adaptation options are lacking; (v) price variability and volatility often threaten food security; and (vi) more attention needs to be paid to the governance of food systems.

5.2 Human security

The concept of human security came to prominence through the 1994 Human Development Report, which defined human security as a "concern with human life and dignity" (UNDP 1994, 22), and which adopted a comprehensive approach by identifying economic, food, health, environmental, personal, community, and political components to human security. In the 21st century three key issues facing humankind are environmental degradation, impoverishment, and the insecurity that can result from either of these two. A review of environment and security work indicates that there is an ongoing need for conceptual and theoretical discussions on the nature of the relationship between environment and security. It is also important to build upon early empirical work that focused on environment and conflict and to provide additional empirical studies on environmental change and its relationship to a broader conception of security. At the same time expanded research networks and improved communication among researchers, policy makers, and NGOs are required in order to develop integrated research projects on environmental change and human security.

5.3 Human health

It is well established that human health is linked to environmental conditions, and that changes in the natural environment may have subtle, or dramatic, effects on health. Timely knowledge of these effects may support our public health infrastructure in devising and implementing strategies to compensate or respond to these effects.

Ecosystems, human health and economy are all sensitive to changes in climate- including both the magnitude and rate of climate change. Climate change is likely to affect human health and well-being through a variety of mechanisms. For example, it can adversely affect the availability of freshwater, food production, and the distribution and seasonal transmission of vector-borne infectious diseases such as malaria, dengue fever and schistosomiasis. The additional stress of climate change will interact in different ways across regions. It can be expected to reduce the ability of some environmental systems to provide, on a sustained basis, key goods and services needed for successful economic and social development, including adequate food, clean air and water, energy, safe shelter and low levels of diseases (IPCC 2001).

In this chapter, current state of knowledge of the associations between weather/climate factors and health outcome(s) for the population(s) concerned, either directly or through multiple pathways is as outlined in Figure 4. The figure shows not only the pathways by which health can be affected by climate change, but also shows the concurrent direct-acting and modifying (conditioning) influences of environmental, social and health system factors.

5.4 Natural disturbances

Changing climate has the potential to increase risks from sea level rise, extreme storm events, and drought. About 25% of the world's population lives with 100 km of the coast. In 2007, the IPCC projected an 18-59 cm sea level rise by 2100, but many scientists argue that this range is too low, and that sea level rise could be as great as 1-6 m (Kopp et al 2009; Jevrejeva *et al.* 2010). People living in low lying regions, such as Bangladesh and Pacific island nations (e.g., Tuvalu and the Maldives) are already experiencing the effects of salt water incursion in their agricultural fields and fresh water supplies. Arctic Inuit communities are battling the loss of coastal villages as a result of increased storm surges from sea level rise.

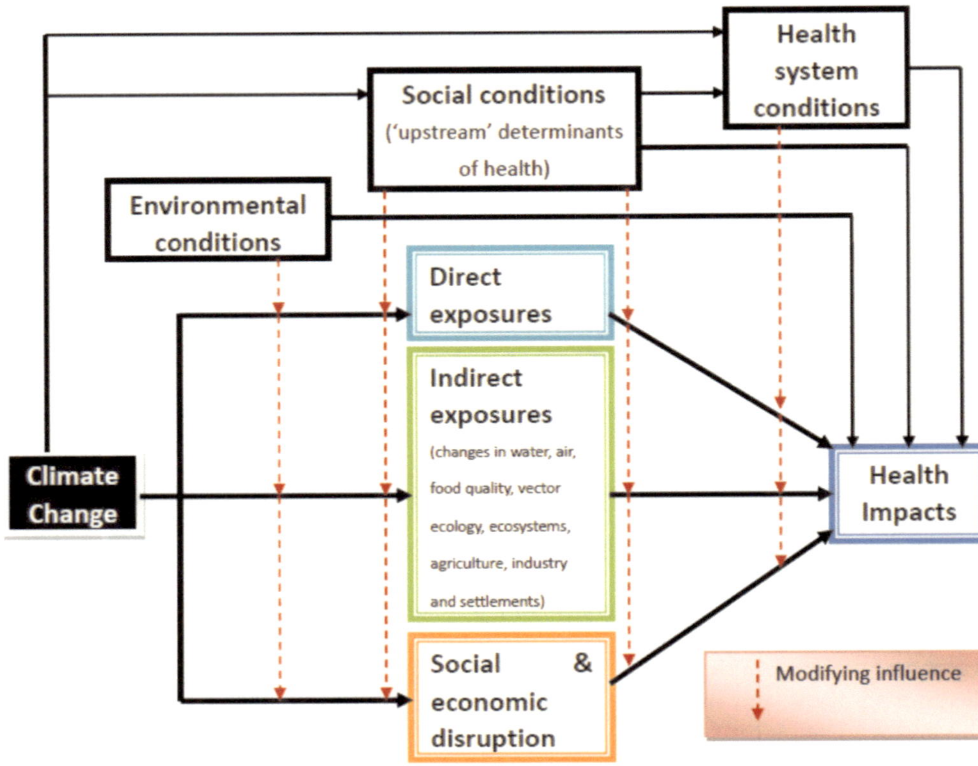

Fig. 4. Schematic diagram of pathways by which climate change affects health, and concurrent direct-acting and modifying (conditioning) influences of environmental, social and health-system factors (Modified from Confalonieri et al 2007).

Extreme precipitation events may become more common in mid-to-high latitude regions, consistent with the prediction that warmer air temperatures, as a result of climate warming, will increase the moisture content of the atmosphere (IPCC 2007). Besides catastrophic flooding and associated property damage, extreme storms are a concern for infrastructure, with cities now faced with the prospect of significant costs associated with roads, dams, and levees being washed out by floods. Regions such as the Sahel are particularly vulnerable to increases in rainfall variability and extremes; the timing of rainfall is as important as the amount of rainfall for agriculture in the Sahel. Extreme rainfall events in semi-arid regions are also likely to lead to increased soil erosion; analysis of interannual variability in rainfall and atmospheric dust content in the Sahel indicates that episodic rainfall events play an important role in generating the erodible material that is necessary for the development of dust storms (Brooks and Legrand, 2001). Atmospheric dust is a major cause of respiratory problems in regions such as the Sahel (Shinn, 2001), and a shift towards higher rainfall variability and intensity might therefore have a negative impact on health.

6. Global response

"The scientific evidence is now overwhelming: climate change presents very serious global risks, and it demands an urgent global response" (Stern Review, 2007). Human perceptions of the natural environment, and the way we use the environment, are socially constructed. The human responses are mitigation and adaptation. However, environmental problems must be addressed from a broader perspective that includes issues of impoverishment and issues of (in) equity, and recognizing the fact that "Space Matters". In the context of global environmental change, it is important to consider the various spatial levels at which both environment and security concerns can be addressed. It is believed that these needs can be best met through an international research program that focuses both on guiding future research and assisting in policy development (at all levels).

Given the severity of global environmental change impacts, various adaptation and mitigation measures are being used in the fight against climate change. The effective targeting of these measures across different sectors such as water, agriculture, tourism, infrastructure development and others requires the use of both practical and innovative strategies. Adaptation measures fall within a broad range, among others, from expanded water harvesting, storage and conservation techniques to the diversification of tourism activities. An example of an adaptation strategy to prevent damage from climate change is shore protection (e.g., dikes, bulkheads, beach nourishment), which can prevent sea level rise from inundating low-lying coastal property, eroding beaches, or worsen flooding. If the costs or environmental impacts of shore protection are high compared with the property being protected, an alternative adaptation strategy would be a planned retreat, in which structures are relocated inland as shores retreat. Also, with current climate fluctuations, a good example of adaptation and coping strategies include farmers planting different crops for different seasons, and wildlife migrating to more suitable habitats as the seasons change.

Mitigation measures include "climate friendly" technological innovations; alternative fuels; sustainable land management practices to name but a few (Nelson et al, 2009). Also, the resilience of biodiversity to climate change can be enhanced by reducing non-climatic stresses in combination with conservation, restoration and sustainable management strategies. Conservation and management strategies that maintain and restore biodiversity can be expected to reduce some of the negative impacts from climate change; however, there are rates and magnitude of climate change for which natural adaptation will become increasingly difficult (CBD 2009).

Furthermore, from the highlights of Stern Review, emerging schemes that allow people to trade reductions in CO_2 have demonstrated that there are many opportunities to cut emissions for less than $25 a tonne. In other words, reducing emissions will make us better off. According to one measure, the benefits over time of actions to shift the world onto a low-carbon path could be in the order of $2.5 trillion each year. The shift to a low-carbon economy will also bring huge opportunities. Markets for low-carbon technologies will be worth at least $500bn, and perhaps much more, by 2050 if the world acts on the scale required. Tackling climate change is the pro-growth strategy; ignoring it will ultimately undermine economic growth (Stern Review, 2007).

Lastly, international policy responses to the global challenges described above have been many and varied. Most of them are based on several UN Conventions such as the UN Conference on Environment and Development, Framework Convention on Climate Change (UNFCC), Convention on Biological Diversity (CBD) or the large number of agreements tied to the UN Convention on the Law of the Sea. At the same time, institutional and organizational weaknesses in some countries and the complex interaction among myriad authorities responsible for ecosystems (including marine and coastal) and environmental management make the implementation of policies a difficult task (Agboola and Braimoh,

2009). Changes in institutional and environmental governance frameworks are sometimes required to create the enabling conditions for effective management of ecosystems, while in other cases existing institutions could meet these needs but face significant implementation barriers (MA, 2005). A lot of commitments to sustainable development have been made at past UN conferences, including Agenda 21 (1992), the Rio Declaration on Environment and Development (1992) and the Johannesburg Plan of Implementation (2002).

In all these, there is the need to examine how far we have come in achieving these commitments before we can channel a better way to moving forward.

7. Conclusion

This chapter dealt on some relevant issues and current dimensions in GEC. It presents certainties as explained by some GEC indicators and predictions, and future uncertainties that may require strategic interventions. In all of these, it foregrounds the need for a broader perspective in tackling the myriads of global environmental challenges. Although, research on the human dimensions of global change concerns human activities that alter the Earth's environment, the driving forces of those activities, the consequences of environmental change for societies and economies, and human responses to the experience or expectation of global change; such research is essential both to understand global change and to inform public policy. This review suggests the need for a more interdisciplinary and integrative perspective on environmental change issues. A more integrated understanding of the complex interactions of human societies and the Earth system is essential if we are to identify vulnerable systems and pursue options that take advantage of opportunities and enhance resilience.

8. References

Agboola, J.I., Braimoh, A.K., (2009) Strategic partnership for sustainable management of aquatic resources. Water Resour. Manage., 23: 2761-2775

Benayas, J. M. R., Newton, A. C., Bullock, J. M., (2009) Enhancement of biodiversity and ecosystem services by ecological restoration: A meta-analysis. Science 325, 1121–1124

Braimoh, A. K., Subramanian, S. M., Elliott, W. S., Gasparatos, A., (2010) Climate and Human-Related Drivers of Biodiversity Decline in Southeast Asia. UNU-IAS Policy Report.

Brooks, N., Legrand, M., (2000) Dust variability over northern Africa and rainfall in the Sahel, in Linking climate change to land surface change, S. McLaren and D. Kniveton (eds.), Dordrecht, Kluwer Academic Publishers, 1-25.

Bruner, A. G., Gullison, R.E., Rice, R. E., da Fonseca, G. A. B., (2001) Effectiveness of parks in protecting tropical biodiversity. Science 291, 125–128.

Butchart, S.H.M., Walpole, M., Collen, B., Strien, A.V., Scharlemann, J.P.W., Almond, R.E.A., Baillie, J.E.M., Bomhard, B., Brown, C., Bruno, J., Carpenter, K.E., Carr, G.M., Chanson, J., Chenery, A.M., Csirke, J., Davidson, N.C., Dentener, F., Foster, M., Galli, A., Galloway, J. N., Genovesi, P., Gregory, R. D., Hockings, M., Kapos, V., Lamarque, J. F., Leverington, F., Loh, J., McGeoch, M.A., McRae, L., Minasyan, A., Morcillo, M. H., Oldfield, T.E.E., Pauly, D., Quader, S., Revenga, C., Sauer, J.R., Skolnik, B., Spear, D., Stanwell-Smith, D., Stuart, S.N., Symes, A., Tierney, M., Tyrrell, T.D., Vie, J.C., Watson, R., (2010) Global biodiversity: indicators of recent declines. Science, 328, 1164–1168.

Caldeira, K., Wickett, M.E. (2003). "Anthropogenic carbon and ocean pH". Nature 425 (6956): 365-365.

Carpenter, S.R., Mooney, H.A., Agard, J., Capistrano, D., DeFries, R.S., Diaz, S., Dietz, T., Duraiappah, A.K., Oteng-Yeboah, A., Pereira, H.M., Perring, C., Reid, W.V., Sarukhan, J., Scholes, R.J., Whyte, A., (2009) Science for managing ecosystem services: beyond the millennium ecosystem assessment. PNAS 106, 1305–1312.

Collins, S. L., Knapp, A. K., Briggs, J. M., Blair, J. M., Steinauer, E. M., (1998) Modulation of diversity by grazing and mowing in native tallgrass prairie. Science 280, 745–747.

Confalonieri, U., Menne, B., Akhtar, R., Ebi, K.L., Hauengue, M., Kovats, R.S., Revich, B., Woodward, A., (2007) Human health. Climate Change 2007: Impacts, Adaptation and Vulnerability. Contribution of Working Group II to the Fourth Assessment Report of the Intergovernmental Panel on Climate Change, M.L. Parry, O.F. Canziani, J.P. Palutikof, P.J. van der Linden and C.E. Hanson, Eds., Cambridge University Press, Cambridge, UK, 391-431.

Convention on Biological Diversity- CBD, (2009) Connecting Biodiversity and Climate Change Mitigation and Adaptation: Report of the Second Ad Hoc Technical Expert Group on Biodiversity and Climate Change. Montreal, Technical Series No. 41, 126 pages

Ebi, K.L., (2011) Climate Change and Health. Encyclopedia of Environmental Health, pp 680-689

Ericksen, P. J., Ingram, J. S. I., Liverman, D.M., (2009) Food security and global environmental change: emerging Challenges. Environmental Science & Policy, 12: 373-377

FAO, (2010) Global Forest Resources Assessment. FAO, Rome, Italy.

Godfray, H.C.J., Beddington, J.R., Crute, I.R., Haddad, L., Lawrence, D., Muir, J.F., Pretty, J., Robinson, S., Thomas, S.M., Toulmin, C. (2010) Food security: the challenge of feeding 9 billion people Science 327, 812-818.

Hall-Spencer, J. M., Rodolfo-Metalpa, R., Martin, S., Ransome, E., Fine, M., Turner, S. M., Rowley, S., Tedesco, D., Buia, M. C., (2008) Volcanic carbon dioxide vents reveal ecosystem effects of ocean acidification. Nature 454: 96-99.

Harley, C. D. G., Hughes, A. R., Hultgren, K. M., Miner, B. G., Sorte, C. J. B., Thornber, C. S., Rodriguez, L. F., Tomanek, L., Williams, S. L., (2006) The impacts of climate change in coastal marine systems. Ecology Letters 9, 228–241

IPCC, (2001) Climate Change 2001: Impacts, Adaptation and Vulnerability. Contribution of Working Group II to the Third Assessment Report of the Intergovernmental Panel on Climate Change. Cambridge, United Kingdom, and New York, United States, Cambridge University Press.

IPCC, (2001) IPCC Special Report on Emissions Scenarios

IPCC, (2007) Core Writing Team; Pachauri, R.K; and Reisinger, A., ed., Climate Change 2007: Synthesis Report, Contribution of Working Groups I, II and III to the Fourth Assessment Report of the Intergovernmental Panel on Climate Change, IPCC, ISBN 92-9169-122-4.

Jager, J., (2002) Global Environmental Change: Human Dimensions. International Encyclopedia of the Social & Behavioral Sciences. pp 6227-6232. DOI:10.1016/B0-08-043076-7/04137-1

Jevrejeva, S., Moore, J. C., Grinsted, A., (2010) How will sea level respond to changes in natural and anthropogenic forcings by 2100? Geophysical Research Letters 37, L07703, 5 pp.

Key, R.M., Kozyr, A., Sabine, C.L., Lee, K., Wanninkhof, R., Bullister, J., Feely, R.A., Millero, F., Mordy, C., Peng, T. H. (2004) "A global ocean carbon climatology: Results from GLODAP". Global Biogeochemical Cycles 18 (4): GB4031.

Kopp, R.E., Simons, F.J., Mitrovica, J.X., Maloof, A.C., and Oppenheimer, M., (2009) Probabilistic assessment of sea level during the last interglacial stage. Nature 462, 863-868

MA, (2005) Millennium Ecosystem Assessment (MA). Ecosystems and Human Well-being: Synthesis. Washington, DC: Island Press.

Marris, E., (2010) UN body will assess ecosystems and biodiversity. Nature 465, 859.

Naeem, S., Bunker, D. E., Hector, A., Loreau, M. & Perrings, C., (2009) Introduction: The ecological and social implications of changing biodiversity. An overview of a decade of biodiversity and ecosystem functioning research. In Biodiversity, Ecosystem Functioning, and Human Wellbeing: An Ecological and Economic Perspective, S. Naeem, D. E. Bunker, A. Hector, M. Loreau & C. Perrings Eds, Oxford University Press, Oxford, United Kingdom, pp. 3-13.

Nelson, G.C., Rosegrant, M.W., Koo, J., Robertson, R., Sulser, T., Zhu, T., Ringler, C., Msangi, S., Palazzo, A., Batka, M., Magalhaes, M., Valmonte-Santos, R., Ewing, M., Lee, D., (2009) Climate Change: Impact on Agriculture and Costs of Adaptation. International Food Policy Research Institute (IFPR).

Orr, J. C., Fabry, V. J., Aumont, O., Bopp, L., Doney, S. C., Feely, R. A., Gnanadesikan, A., Gruber, N., Ishida, A., Joos, F., Key, R. M., Lindsay, K., Maier-Reimer, E., Matear, R., Monfray, P., Mouchet, A., Najjar, R.G., Plattner, G., Rodgers, K. B., Sabine, C. L., Sarmiento, J. L., Schlitzer, R., Slater, R. D., Totterdell, I. J., Weirig, M., Yamanaka, Y., Yool, A., (2005) "Anthropogenic ocean acidification over the twenty-first century and its impact on calcifying organisms". Nature 437 (7059): 681–686.

Pidwirny, M., (2006) "The Hydrologic Cycle". Fundamentals of Physical Geography, 2nd Edition. Date Viewed: 20/09/2011. http://www.physicalgeography.net/fundamentals/8b.html

Sala, O.E., Chapin III, F.S., Armesto, J.J., Berlow, E., Bloomfield, J., Dirzo, R., Huber-Sannwald, E., Huennecke, L.F., Jackson, R.B., Kinzig, A., Leemans, R., Lodge, D.M., Mooney, H.A., Oesterheld, M., Poff, N.L., Sykes, M.T., Walker, B., Walker, M., Wall, D.H., (2000) Global biodiversity scenarios for the year 2100. Science, 287, 1770–1774.

Sala, O. E., Meyerson, L. A., Parmesan, C., (2009) Biodiversity change and human health: from ecosystem services to spread of disease. Island Press. pp. 3–5. ISBN 9781597264976. Retrieved 28 June 2011.

Shinn, Eugene A., (2001) African dust causes widespread environmental distress, Environmental Geology

Steffen, W., Sanderson, A., Tyson, P.D., Jager, J., Matson, P.A., Moore, III, B., Oldfield, F., Richardson, K., Schellnhuber, H.J., Turner, II, B.L., Wasson, R.J. (Eds.), 2003. Global Change and the Earth System: A Planet Under Pressure. SpringerVerlag, Berlin/New York.

Stern, N., (2007) The Economics of Climate Change. The Stern Review. Cambridge University Press, Cambridge.

UNFCC, (2003) United Nations Framework Convention on Climate Change. Bonn UNFCCC 2003

UNUDP, (1994) Human Development Report 1994. New York Oxford, Oxford University Press

Vitousek, P. M., (1994) Beyond Global Warming: Ecology and Global Change. Ecology 75:1861–1876. [doi: 10.2307/1941591]

Woodward, J. and Buckingham, S., (2008) 'Global climate change'. In Buckingham, S. and Turner, M. (eds.) Understanding environmental issues. London; Sage, pp. 175-206

WRI., (2006) Climate Analysis Indicators Tool (CAIT) on-line database version 3.0. World Resources Institute, Washington DC. Available at <www.cait.wri.org> [Accessed 15 March 2009]

2

Relevant Issues for Sustainable Agriculture in Sub-Saharan Africa

Ademola A. Adenle[1] and Julius I. Agboola[2]
[1]*United Nations University-Institute of Advanced Studies, Yokohama, 6F International Organizations Center, Pacifico-Yokohama*
[2]*United Nations University, Institute of Advanced Studies, Operating Unit in Ishikawa/Kanazawa, Kanazawa, Ishikawa, Japan*

1. Introduction

Given the food security problems, rising population and unsustainable agriculture practice through conventional approach, the role of agricultural sustainability can not be overemphasised. But not many countries in Africa especially Sub-Saharan African (SSA) are practising sustainable agriculture, even if there are, they are very few and uncoordinated. Agricultural sustainability has a key role to play towards agricultural development in SSA, particularly improving agricultural productivity and food security, conserving biodiversity and creating friendly environment.

This chapter focuses on relevant issues including constraints that affect sustainable agricultural development in SSA. In section two, we review relevant issues for sustainable agricultural development such as land use system, government institutions and infrastructure and agricultural technology, demonstrating current literature examples in our arguments. Third section focuses on effect of climate change on agricultural production, also exploring the link between climate change and agriculture biodiversity as well as climate regimes in SSA. Fourth section examines conservation of biodiversity which is also relevant to sustainable agricultural development, particularly when considering the roles of ecosystem and ecological species in the environments. The chapter concludes by summarising the policy implication that requires the attention of national governments and relevant organisations towards successful adoption of sustainable agriculture practice in SSA.

2. Relevant issues for sustainable agricultural development

2.1 Land use system

Land use in Africa has undergone evolutionary transformation from the simple hunting to more complex sedentary, shifting or commercial cultivation system. Rural land use has evolved from bush hunting in tropical Africa like Kalahari Desert and nomads practice like Pygmies in the Zaire/Congo Basins going through shifting cultivation that is widely practiced in the Miombo woodlands of SSA where the soil and vegetation are allowed to fallow before further cultivation (Pritchard, 1979). The sedentary or permanent agriculture is known to be practiced among Souafa population in the Sahara where vegetation, fruit, palm tree, eucalyptus trees and rice are planted, and animals such as goats, camels and sheep are reared. In East Africa (Uganda) the root and tubers, cereals and legumes also form agronomic practices of sedentary agriculture. The population engaged in this kind of farming practices are known to settle in the villages in many parts of Africa but Rwandans farmers settled across tops of the hills with fertile soils where bananas are planted around the houses, and coffee grown a bit farther down the hill (Clay and Lewis, 1990). This gradually evolves to shifting cultivation where farmers can move from one land to another without the right to possession of land after vacation, and the use of axe and hoe are prevalent for cultivating land particularly in high rainfall regions. Under this indigenous agronomic system and application of low level of agricultural technology, the land is put into use to ensure food security and sustainable agricultural practices in the communities. Following this agronomic system, particularly during the colonial era in some of the African countries like Zambia, Kenya, Zimbabwe and Malawi, a semi-permanent, ox and tractor-plough cultivation emerge where arable land can be put into use for as long as 10 years before the land is allowed to fallow for a limited period of time coupled with the use of cattle manure for maintenance of soil fertility (Trapnell and Clothier, 1996). During this period, cash crops such as tea, maize, tobacco, cotton and others were largely cultivated for commercial purposes which resulted to a combination of sedentary, modern and commercial cultivation as introduced from colonial masters to African farmers.

Agricultural practice then was well coordinated and more sustainable compared to African present situation. The traditional system of using slopes, for examples, wood lots, fallow and pasture (Clay and Lewis, 1990), and plant cover such as mulches and crops offered effective protection against erosion (Wischmeier and Smith, 1978). The traditional land use system was not under increasing pressure and competition. For example, in Zambia and Ethiopia, the process of acquiring land for farming purposes was duly followed as stipulated in the law (Gilks, 1975; White, 1959). After several African countries gained political independence and as the population increased, modern agricultural techniques were diffused leading to the birth of an indigenous group of commercial farmers (Baylies, 1979) and more individual traditional agricultural practice emerged, and as a result the land tenure law system expressed in English law was transformed and revised in response to social, cultural, political and economic changes in these countries (Mvunga, 1980). But traditional agricultural practice in Africa that involves true shifting cultivation is now rare, pressure of increasing population due to reduction of the fallow has led to massive deterioration in food production, and current land tenure system constrains sustainable agricultural practices. In majority of African countries, national policies on land and economic development are not aligned with agricultural practices, particularly among the rural poor.

The era of land degradation and insecure land tenure system; lack of formal documentation as to who has the right to the land remains an obstacle to improve agricultural productivity and food security, to promote investment and encourage better natural-resource management, thus contributing to increasing level of poverty in Africa. Land degradation has become one of major problems with nearly 2 billion hectares of land affected in Africa, and 300 million hectares out of this accounts for considerable loss of nutrients, desertification and soil erosion (Nkonya et al., 2008; Pintstrup-Anderson and Pandya-lorch, 1995). Loss of nutrients and soil erosion are basic indicators of land degradation. For example, in Uganda, it was reported that six major agro-ecological and farming zones were affected due to soil depletion with an average of 179 kg/ha of N, P, and K per year equivalent to about 1.2 percent of the nutrient stock stored in the topsoil (0–20 cm depth). Nkonya et al (2008) argue further that the replacement of depleted nutrients is equivalent to one-fifth of the household income obtainable from agricultural production based on minimum price of inorganic fertilizers. Much of highlands of eastern and central Africa are affected, particularly the densely populated areas and step mountain slopes with volcanic fertile soils (Henao and Baanante, 2006; Smaling et al., 1997; Voortman et al., 2000). For example, in Ethiopia, around 30% of agriculture was degraded in 1990 due to soil erosion that occurred in 1970s which was equivalent to10 billion metric tons per year of soil (Hutchinson et al., 1996; Myers, 1986). A more recent report has shown that annual agriculture gross domestic product (GDP) of Africa can be reduced by 3% due to land degradation (Jansky and Chandran, 2004).

Africa exploding population leading to more people in urban area than rural, and more occupation of land in urban areas without proper legal structure for land ownership has become a big threat to economy, innovation and job opportunity in the city. Africa rural population has increased by 265% and urban population with nine fold increase between 1950 and 2000 (UN, 2004). And large dependence of rural population on subsistence agriculture for their livelihoods due to population pressure invariably result to expansion of agricultural lands (Wood et al., 2000). Given the current land tenure system, improving agricultural productivity and food security that will contribute to economy growth and sustainable development still remain a big challenge in Africa. But the question is how to create enabling environment to safeguard the livelihood of small-scale farmers that account for over 80% of agricultural production in rural Africa to have equal right or access to land, particularly enacting law to support land titling and registration that can benefit poorer society. Also, in urban areas, how can hundreds of millions of people who live in crudely built shacks have equal right to land? As over 72% of Africa urban population live in the slum areas (UN-HABITAT, 2003). It has been reported that farmers near urban areas can be easily displaced due to rising land values and their agricultural land being converted to building for public works or commercial purposes either legally or illegally in Africa (Benjaminsen and Sjaastad, 2002).

While in many countries, land or private assets acquired by governments must be expropriated as stipulated under country constitution, for example, the United States (US) constitution requires compensation for all takings of private property, and the Philippine and Brazil constitution similarly requires that payment through compensation, and Cambodia constitution mandates that the states make fair and just compensation for taking possession of land from any person (ECV, 2011). Loss of land or private assets acquired is often not compensated in Africa, although payment of improvement alone can be enforced in many countries but often inadequate and late (Benjaminsen and Sjaastad, 2008; Cotula, 2008; Kasanga and Kotey, 2001). Kasanga and Kotey (2001) demonstrate that Ghanaian governments have not yet paid compensation for land taken which was estimated to be billions of Ghanaian Cedis. Acquiring property and land by the governments without compensation can be more devastating in other African countries, particularly among the poorer and less politically power groups. For example, in Zimbabwe, almost a quarter of a million people who had an illegal markets and business activities near capital Harare and other urban centres in 2005 lost their properties, lands and homes as mass eviction was carried out by the government without compensation, and this left majority of these people in a permanent difficult situation for the rest of their lives (de Plessis, 2005).

2.2 Government institutions and infrastructures

The role of government institutions is fundamental to sustainable agricultural development in Africa. In modernised society, government institution plays a critical role in information-dissemination, interaction, knowledge-sharing, awareness-creation and public education, particularly in relation to development and implementation of agricultural policies towards sustainable agriculture. Switzerland is a good example, recent interviews with key stakeholders demonstrates that agriculture practice is quite sustainable and government institution has been very effective, although adaptive approach for sustainable agricultural practices dated back to 19th century in the country has changed after World Wars in Europe (Aerni, 2009). A broad-based perspective on sustainable agriculture also emphasizes on components such as the roles of diverse actors, linkages, innovations, development strategies, partnerships and networks across other institutions to ensure better delivery packages. All these components are integrated together to allow present needs to be met without compromising future needs which is one of the hallmarks of a sustainable agricultural system.

Large majority of African populations engaged in agricultural practices is mainly found in rural areas but often not involved in active participation of debates or issues relating to agricultural development in the rural communities. Common understanding on introduction of new technology or any concept for agricultural development, and clarity on the roles of different stakeholders including farmer groups is very vital. While the responsibility of government institutions, particularly the Ministry of Agriculture and other relevant agencies at national level is to carry along and work in partnerships with local governments that deal directly with local communities, it is the government that makes the decisions, for example, the kind of new technology that suit the farmers and their environments without prior consultations. A report by (Uphoff, 1986) describes that local governments are important for mobilising resources and regulating their use with location specific knowledge so as to produce a locally interpreted and oriented results that can suit local people and their environments with a view to facilitating sustainable agriculture and rural development. He describes further that using participatory approach such as "bottom up" that begins with self-help efforts and later engages higher-level resources with limited efforts or commitments is inefficient but proposes a frame work of local institutions based on demand-driven strategy and result-oriented. A framework based on demand-driven is common to majority of agricultural practices in African countries, but however African governments often fail to fulfil their tasks and responsibility in this regard. For example, in Kenya, current agricultural extension services are demand-driven and require adequate capital to cover transport and other costs but extension officers have little or nothing to render their services (KSP, 2005; KSRA, 2004).

Without doubt, efficient and effective agricultural extension is one of the keys to increasing agricultural production but unfortunately agricultural extension services in majority of African countries have been alarmingly poor. For example, the most five important difficulties as stated by Nigerian extension officers in five states in order of magnitude are given as follows; insufficient transport facilities, low prices and lack of proper markets, lack of cooperation from other agencies in program implementation, lack of staff motivation and inadequate technical training in agriculture (Nigel, 1989). In addition, Tanzanian extension officers also stated and rated the following factors for their inability to deliver adequate and necessary training to the farmers; lack of transportation (50%) unavailability of inputs and training facilities (49%), lack of research input and technical support (43%), lack of incentives and administrative support (31%). These factors remain a huge problem to render adequate extension services to the poor farmers, particularly in the rural areas and can not be remedied by extension officer's individual effort which indicates failure in government institutions. However, non-governmental organisations (NGOs) have been playing an increasingly important role (e.g. research and extension services) in working with African local communities, particularly where huge holes have been created by country governments with a view to closing the gaps and contribute to sustainable agricultural developments in Africa (Levine, 2002; Rukuni et al., 1998). For examples, in Southern Africa and Eastern Africa, farmers are being assisted by forming a strong local group to shift from conventional to sustainable agriculture, integrating people into decision-making process, revitalising farmers' organisations as well as improving markets links through holistic approach. It was estimated by World Bank that over $7.6 billion of aid have been distributed through international NGOs to developing countries including Africa continent in 1992 alone (WB, 2002), one can imagine what could have been spent between that period of time and now.

Lack of well-developed and maintained infrastructures has been an obstacle to agricultural growth and rural development in Africa continent for many decades. The poor state of communication, transportation, irrigation and storage facilities across different African countries does not encourage sustainable agricultural development. Poor infrastructure affects the quality and frequency of services, for example, it prevents access to important services such as extension, market information, credits, health and education in Africa. Access to information and knowledge are important factors to speed up agricultural development, particularly in adoption of improved agricultural practices, marketing system and effective post-harvest management (Bertolini, 2004; Poole and Kenny, 2003). But majority of African farmers lack access to the right information and knowledge on good agricultural practices due to inability of the governments to respond to changing needs of farming communities. Most roads in rural areas are not motorable, and where rural roads exist, they are often poorly maintained and later abandoned. According to (Riverson et al., 1991), the maintenance standards of the rural roads in SSA have fallen significantly during the 1980s, for example, 42% of unpaved roads were reported to be in poor conditions in 1988 compared to 28% in 1984. Due to unreliable rainfall in many African countries, farming practices can be extremely difficult and irrigation can be expensive for the small-scale farmers, and there is lack of well-developed and locally appropriate means of small-scale irrigation that can benefit the farmers. Also, lack of access to credit facilities means farmers can not afford to buy agricultural inputs such as fertilizers and seeds, and micro-credit schemes that could support rural finance are rare and not widely promoted. Even when they are available, they are inefficient, inconsistent and unreliable. Moreover, African agricultural products lack competiveness and access to global trade and international markets that determine economic growth and development prospect of any particular country. African farmers often find it difficult to meet quality standard required to export their agricultural products due to lack of skills to invest in marketing, capital and processing facilities that can increase and add values to their final market output. As a result of bad policy, market failure and very limited mechanized farming system coupled with poor infrastructure and dispersed settlements, most private sectors as demonstrated in Tanzania are unwilling to invest (Ponte, 2001) and leaving them in an unfavourable and largely uninfluential position in the world trading system (Allen and Thompson, 1997; Stiglitz, 2003). Taken together, agricultural production system in Africa is characterised by weak linkages (supply chains, firm structure, organisational support, extension services and poor infrastructure) and poor systemic coordination.

2.3 Agricultural technology

A wide variety of technology is always associated with sustainable agricultural development, particularly with focus on the kind of technology that will be of a greater advantage to developing countries including African continent where agriculture is the main escape route from the poverty. The introduction of new agricultural technologies into African farming system is important for the region development. Given high level of poverty, food insecurity, rising population, pattern of global climate and low yields, adoption of improved technology is fundamental to solving some of these agricultural problems. A recent port on agricultural technology use in Mozambique shows that agricultural productivity is generally low, for example, maize yields are estimated at 1.4 tons/ha which is far below the potential yields of 5 - 6.5 tons/ha (Zavale et al., 2006). However, a good number of literatures have shown that adoption of improved agricultural technologies can lead to higher crop yields, lower food prices, higher real wages for unskilled workers, improved food security and reduced poverty (Cunguara and Darnhofer, 2011; Karanja et al., 2003; Kassie et al., 2011; Kijima et al., 2008; Minten and Barret, 2008). Kassie *et al* (2011) reports that adoption of improved groundnut varieties in Uganda, has a great potential to increase house income in the range of US$130-US$254 leading to reduced level of poverty incidence. According to Minten and Barret (2008), they conclude that improved agricultural technology diffusion appears to be the most effective means of improving agricultural productivity, lowering poverty rates and food security problems in Madagascar. They conclude further that, other than these means, there are no magic bullets to food production and for better agricultural performance in Africa. Despite the fact that, new agricultural technologies can bring a lot of advantages to the farmers, and as observed by (Zavale et al., 2006); the use of improved agricultural technologies is very limited and unequal, and most of agricultural production is rainfed which enhances the effectiveness of improved agricultural technologies. The observation by Zavale and his colleagues is consistent with (Cunguara and Darnhofer, 2011) that describe that an adequate rainfall is required for most improved technologies to be productive and effective. In view of the fact that productivity enhanced technology is required for sustainable agriculture around world (especially developing countries) and the need to increase agricultural production is important, there is no major consensus yet as to technological innovation that can deliver the result while considering factors such as erratic weather, inconsistent policy regime and poor infrastructures in developing countries (DFID, 2008). However, there is overwhelming response and support for research on increasing agricultural productivity in developing countries. For example, greater emphasis is laid on research based on the use of simple and low-cost technology including applying indigenous knowledge for agricultural practices and agricultural biotechnology (e.g. traditional plant breeding and genetic modification) (DFID, 2008; UNCTAD-UNEP, 2008).

3. Climate change in the Sub-Saharan Africa

According to the Intergovernmental Panel on Climate Change (IPCC) report, Africa is the most vulnerable continent to climate change even though its contribution to the increasing concentrations of greenhouse gases has been minimal. Sub-Saharan African countries are particularly vulnerable to climate change because of their dependence on rainfed agriculture, high levels of poverty, low levels of human and physical capital, and poor infrastructure (Nelson et al., 2009). Also, due to its high exposure to climate change impact and as widespread poverty severely limits its capabilities to adapt; it is one of the most adversely affected regions in the world. Thus, there is a wide consensus that climate change will worsen food security in Africa through continuous climatic shifts, as well as an increase in extreme events.

With 40 percent of its population living on arid, semi-arid, or dry sub-humid areas (UNDP 1997), Africa is one of the areas of the world most exposed to global warming. It has experienced a warming of approximately 0.7°C during the past century, and the temperature is expected to increase by between 0.2°C and 0.5°C each decade. Moreover, the decline in rainfall observed in the Sahel in the last 25 years was the most substantial and sustained recorded anywhere in the world since instrumental measurement began (Hulme and Kelly, 1993). In these contexts, climate change means that future farming and food systems will face substantially modified environments as they struggle to meet the demands of a changing global population. Efforts to cope with the stresses caused by growth in demand for food and water will be confounded by a range of stresses, for example, higher temperatures, changing rainfall patterns and rising sea levels.

Climate change increases the risk of reductions in crop and livestock yields. Within a given region, different crops and livestock are subject to different degrees of impacts from current and projected climate change (Lobel et al., 2008). The negative effects of climate change on crop production are especially pronounced in SSA, as the agriculture sector accounts for a large share of GDP, export earnings, and employment in most African countries (Nelson et al., 2009), and the vast majority of the poor reside in rural areas and depend on agriculture for their livelihoods. The numbers of hydro-meteorological disaster including extreme temperature, from a large scale to a small scale country level, continue to provide indicators of a changing climate as shown in Table 1 (Guha-Sapir et al., 2011), regional analysis on natural disaster occurrence and impacts revealed that in 2010, Africa experienced slightly more disasters (69) compared to the annual average disaster occurrence during the last decade (64). This was mostly due to an increase in the number of hydrological disasters (events caused by deviations in the normal water cycle and/or overflow of bodies of water caused by wind set-up). On average, other disasters occurred less frequently than observed over the last decade. Consequently, hydrological disasters took an 82.6% share in 2010, while from 2000 to 2009 they represented 66.5% of all disasters in Africa.

Number of Natural Disaster	**Africa**	**Global**	**Africa's Global percentage**
Climatological 2010	6	50	12
Avg. 2000-9	9	54	16
Geophysical 2010	1	31	3.2
Avg. 2000-9	3	31	9.7
Hydrological 2010	57	216	26
Avg. 2000-9	43	192	22
Meteorological 2010	5	88	5.7
Avg. 2000-9	9	105	8.6
Total 2010	69	385	18
Avg. 2000-9	64	387	16.5

Table 1. Regional analysis on natural disaster occurrence and impacts (Adapted from Guha-Sapir et al 2010).

3.1 Linking climate change and agricultural biodiversity

The issues of climate change and biodiversity are interconnected, not only through climate change effects on biodiversity, but also through changes in biodiversity that affect climate change (CBD, 2009). The work of IPCC has made us all aware that Climate Change is likely to be the main driver of biodiversity loss in the future, suggesting a strong nexus between climate change and agricultural biodiversity. Biodiversity has already been affected by recent climate change and projected climate change for the 21st century is expected to affect all aspects of biodiversity (IPCC, 2002). Also, in like manner, Isbell (2010) hypothesized that Global ecosystem changes are currently destabilizing species interactions (Figure 1) and that this will lead to future declines in biodiversity, ecosystem functioning, and ecosystem stability. This means we must start radically reducing emissions now and stay on a low emissions pathway to avoid increasing the amount of CO_2 in the atmosphere.

However, in response to these challenges, the Consultative Group on International Agricultural Research (CGIAR) Research Program on Climate Change, Agriculture and Food Security (CCAFS) research initiative launched by the CGIAR and the Earth System Science Partnership (ESSP) has recognized the need to overcome the threats to agriculture and food security in a changing climate, exploring new ways of helping vulnerable rural communities adjust to global changes in climate. One major way suggested was to address the increasing challenge of global warming and declining food security on agricultural practices, policies and measures through a strategic collaboration.

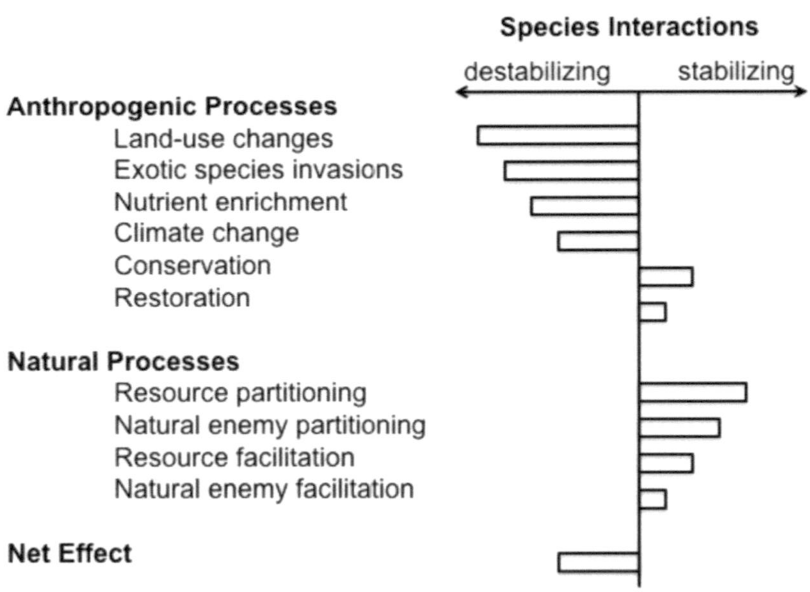

Fig. 1. Hypothesized effects of anthropogenic and natural processes on stabilizing species interactions. Source: (Isbell, 2010)

Here, we opine that the relationship between climate Change and agricultural biodiversity should be viewed in a synergistic approach that assumes that sustainable environmental management and sustainable agriculture can be achieved globally and especially in the SSA. The links between biodiversity and climate change flow both ways (CBD, 2009).

Agricultural Biodiversity, and associated ecosystem services are the cornerstone of sustainable development including food security. Biodiversity also has a very important role to play in climate change mitigation and adaptation. Whatever agricultural revolution Africa wants to undertake, it needs to take into account the carrying capacity of the natural resources and adapt to it.

3.2 Climate regimes and zones

SSA has a wide variety of climate regimes and zones (Fig. 2). The most common regimes are the tropical wet and dry, tropical wet, and tropical dry. In the tropical wet-and-dry climate, there is a distinct dry season during the winter months characterized with droughts and less rainfall. Rainfall occurs during the remainder of the year and can be highly irregular, varying tremendously from one year to the next. In this climate regime, destructive floods can be followed by severe drought. Vegetation in the tropical wet-and-dry regions of SSA is mainly tall savanna grass and low, drought resistant deciduous trees. In this climate regime, sustainable agricultural activities is seasonal and widely considered to be the zone at greatest risk of declining agricultural production at present, and parts of it have been severely affected by drought and food shortages in recent years. At the same time, the parts of the savanna that receive adequate rain have an enormous potential for the expansion of rainfed agricultural production (Higgins et al. 1982).

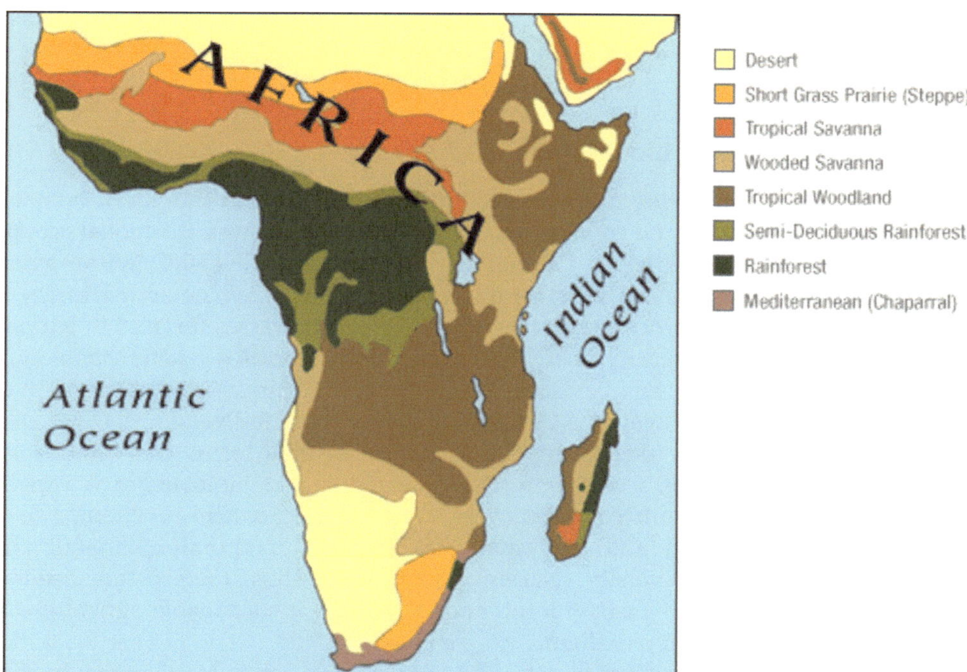

Fig. 2. Climatic zones or biomes in SSA. Source: NASA Earth Observatory.

The tropical wet climate regime affects much of western central Africa. In this region, trade winds from the northern and southern hemisphere meet. As the winds meet, air is forced upward. As the air rises, moisture condenses and towering rain-producing clouds, also called cumulonimbus clouds, form. These cumulonimbus clouds produce heavy rainfall. The heavy rainfall and warm temperatures provide the conditions necessary for the growth of tropical rain forest. Rain forests have higher plant diversity than any other habitat on earth. In terms of the amount of biomass, the natural vegetation is very luxuriant in tropical rainforests, but the natural conditions for developing farming in tropical rainforests are not so favourable. A vivid example is the intensive subsistence cultivation and small holdings which is practised in Nigeria or Ghana in West Africa (i.e. wet rice cultivation). Agricultural use of some rainforest land proves to be a failure because of the nutrient-deficient, acidic soils of these forests. Nevertheless, many commercial agricultural projects are still carried out on rainforest lands, although many of these revert to cattle pasture after soils are depleted.

Tropical dry climates are characterized by very little rainfall and high temperatures. Even when rain does fall, the region remains dry because the high temperatures cause high rates of evaporation. The tropical dry climate can be divided into the semi-arid and arid climates. The semi-arid regions receive more rainfall than the arid regions. Vegetation in the semi-arid region is mostly short grass prairie. Arid environments receive very little rainfall and are characterized as deserts. Land use in semiarid agriculture is characterized by subsistence agriculture and nomadic pastoralism. Because livestock is considered an important component in the livelihoods of semiarid communities, degradation has always been attributed to this sub sector (Sidahmed and Yazman, 1994). According to the World Resources Institute (WRI, 1992), over grazing is the pervasive cause of soil degradation. It has been estimated that overgrazing causes land degradation of 49 % in semiarid regions of Africa.

4. Conserving agricultural biodiversity

Biodiversity is the diversity of life on Earth and includes the richness (number), evenness (equity of relative abundance), and composition (types) of species, alleles, functional groups, or ecosystems. Agricultural biodiversity, sometimes called Agrobiodiversity, "encompasses the variety and variability of animals, plants and micro-organisms which are necessary to sustain key functions of the agroecosystem, its structure and processes for, and in support of, food production and food security"(FAO, 1998). It further "comprises genetic, population, species, community, ecosystem, and landscape components and human interactions with all these''(Jackson et al., 2005). However, MA (2005) suggests that biodiversity declines may diminish human wellbeing by decreasing the services that ecosystems can provide for people (Figure 3). A group of scientists further postulated that biodiversity is rapidly declining worldwide, and there is considerable evidence that ecosystem functioning (e.g., productivity, nutrient cycling) and ecosystem stability (i.e., temporal invariability of productivity) depend on biodiversity (Naeem et al., 2009). Thus, there is the need to formulate and implement biodiversity conservation strategies if sustainable agriculture in SSA is to become a development paradigm.

In terrestrial habitats, tropical regions like the SSA are typically rich in biodiversity. In SSA, agriculture represents 20% to 30% of GDP and 50% of exports. In some cases, 60% to 90% of the labour forces are employed in agriculture (Peter, 2011). Most agricultural activity is subsistence farming. This has made agricultural activity vulnerable to climate change and global warming. Biotechnology has been advocated to create high yield, pest and environmentally resistant crops in the hands of small farmers. As a result, maintaining agrobiodiversity may be critical in developing climate-change-resistant crop and livestock varieties and genotypes, such as those resistant to drought, heat stress, disease, and saline conditions (Fowler, 2008; Kotschi, 2007), and to ensure the continued survival of crop wild relatives (Jarvis et al., 2008). Also, given the above-mentioned impacts of climate change on agricultural systems, practices that enhance soil conservation and sustainable use and maintain favorable microclimates are important for adaptation in agriculture.

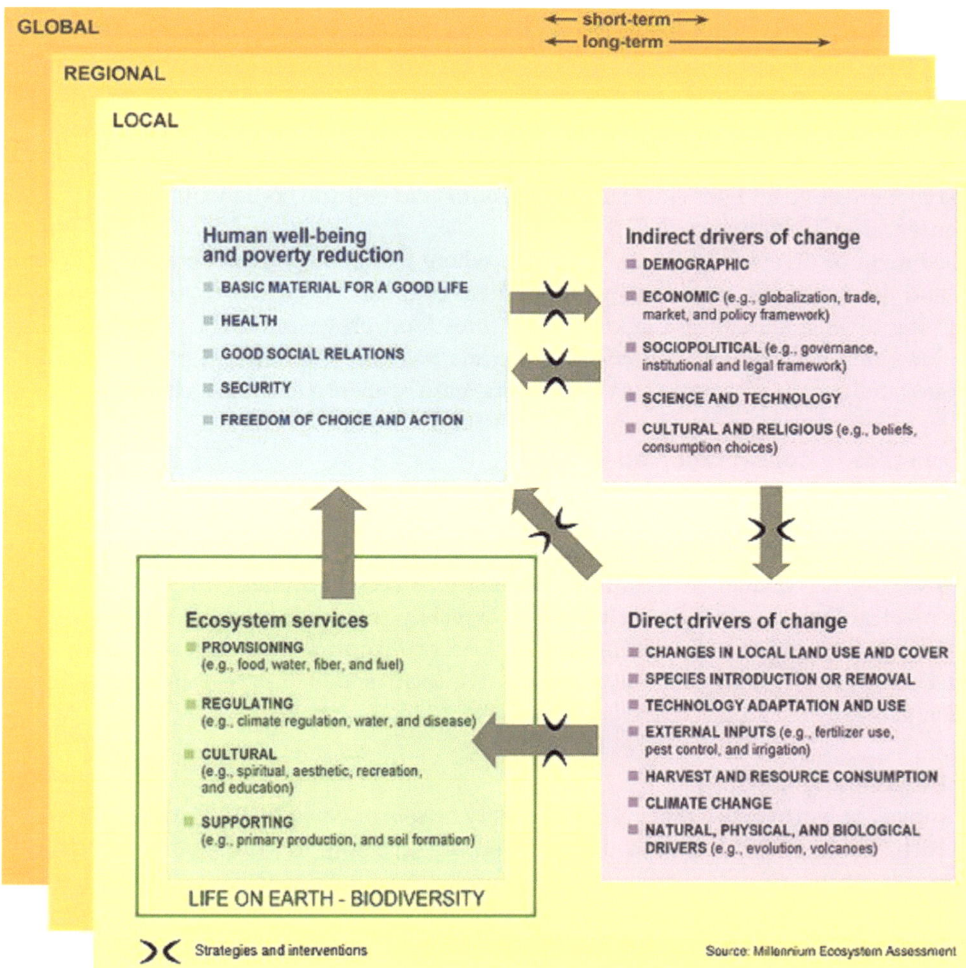

Fig. 3. A schematic image illustrating the relationship between biodiversity, ecosystem services, human well-being, and poverty (MA, 2005). The illustration shows where conservation action, strategies and plans can influence the drivers of the current biodiversity crisis at local, regional, to global scales.

Conserving crop and livestock diversity in many cases helps maintain local knowledge concerning management and use. Developing new varieties may, in addition to meeting adaptation needs, generate co-benefits in the context of health and biodiversity conservation and sustainable use. For example, varieties resistant to crop diseases may contribute to the reduction of pesticide use, also, the use of currently under-utilized crops and livestock can help to maintain diverse and more stable agroecosystems (Bowe, 2007). Also, cultivated and traditional crop varieties represent high levels of genetic diversity and are therefore the focus of most crop genetic resources conservation efforts. Agricultural biodiversity is the basis of our agricultural food chain, developed and safeguarded by farmers, livestock breeders, forest workers, fishermen and indigenous peoples throughout the world. The conservation ethic advocates management of natural resources for the purpose of sustaining biodiversity in species, ecosystems, the evolutionary process, and human culture and society (Dyke, 2008; Wake and Vredenburg, 2008). Thus, the use of agricultural biodiversity (as opposed to non diverse production methods) can contribute to food security and livelihood security. In conserving agricultural biodiversity, ecosystem-based adaptation, which integrates the use of biodiversity and ecosystem services into an overall adaptation strategy, can be cost-effective and generate social, economic and cultural co-benefits and contribute to the conservation of biodiversity (CBD, 2009).

Establishment of diverse agricultural systems, where using indigenous knowledge of specific crop and livestock varieties, maintaining genetic diversity of crops and livestock, and conserving diverse agricultural landscapes secures food provision in changing local climatic conditions; and establishing and effectively managing protected-area systems to ensure the continued delivery of ecosystem services that increase resilience to climate change.

4.1 Examples of conservation approaches

4.1.1 Resource allocation

This requires focusing on limited areas of higher potential biodiversity promises greater immediate return on investment than spreading resources evenly or focusing on areas of little diversity but greater interest in biodiversity. A second strategy focuses on areas that retain most of their original diversity, which typically require little or no restoration. These are typically non-urbanized, non-agricultural areas. Tropical areas often fit both criteria, given their natively high diversity and relative lack of development (Jones-Walters and Mulder, 2009).

4.1.2 Biodiversity banking

Also known as biodiversity trading, biodiversity offsets or conservation banking is a process by which biodiversity loss can be reduced by creating a framework which allows biodiversity to be reliably measured, and market based solutions applied to improving biodiversity. Biodiversity banking provides a means to place a monetary value on ecosystem services. One example is the Australian Native Vegetation Management Framework.

4.1.3 Gene banks

Genes are responsible for the traits exhibited by organisms and, as populations of species decrease in size or go extinct, unique genetic variants are lost. Thus, a gene bank helps preserve genetic material, be it plant or animal. In plants, this could be by freezing cuts from the plant, or stocking the seeds. In animals, this is the freezing of sperm and eggs in zoological freezers until further need. In plants, it is possible to unfreeze the material and propagate it, however, in animals; a living female is required for artificial insemination. While it is often difficult to utilize frozen animal sperm and eggs, there are many examples of it being done successfully. In an effort to conserve agricultural biodiversity, gene banks are used to store and conserve the plant genetic resources of major crop plants and their crop wild relatives. There are many gene banks all over the world, with the Svalbard Global Seed Vault being probably the most famous one.

Another approach that may be considered is: reducing and better targeting of pesticides allows more species to survive in agricultural and urbanized areas.

Lastly, today's loss of biodiversity has been primarily adduced to habitat alteration caused by human activities, thus, halting or reducing the current spate in human adverse activities amongst others will be attempt to conservation of agricultural biodiversity.

5. Conclusion

This chapter has examined issues that are relevant and at the same time important for sustainable agricultural development in SSA. The issues examined reflect many constraints that are still facing sustainable agriculture practice in SSA, thus demanding serious attention from different actors, particularly the governments, private sector and international organisations. For example, land management, infrastructures, access to appropriate technology and vulnerability to climate change still remain a big challenge for sustainable agricultural development in majority of African countries. Given the challenges and constraints facing the agriculture in this region, proactive steps and concerted efforts as well as creating enabling policy environment must be put in place by the country government towards solving these problems. Individual country government, particularly the Ministry of Agriculture should introduce and integrate policy based on sustainable agricultural options into national targets, the extension programme and agricultural curriculum. Government policies should promote sustainable and conservation agriculture practices with less reliance on conventional farming.

Low cost-agriculture technology based on traditional knowledge should form part of the government policies so as to facilitate and encourage sustainable agriculture among small-scale farmers. Government policies should embrace new technological innovation like genetic modification technology that has great potentials to produce disease resistant crops, drought tolerant crops, high-yielding varieties crops as well as improving quality of life for sustainable development. Investing in agricultural research and development and training of extension workers will facilitate sustainable agriculture. Active participation and engagement of farmers that meet the public interest particularly in the development of new technology should be clearly spelt out in government policy. Improved access to higher-value markets, agricultural inputs and good infrastructures such as roads, transport, storage facilities and effective communication system would motivate many more farmers to engage in sustainable agriculture practices. Land reform policy and measures should be provided to aid and strengthen management of agricultural resources and protect the ecological environment. In order to tackle climate change problems, government policy should focus on best practices identified for enhanced agricultural productivity, increased resilience and low carbon agriculture, and environmental sustainability. Hardly any African country has integrated sustainable agriculture practise into its agricultural policies, and adopting sustainable agriculture and integrating into national policies will go a long way towards solving food security problems, combating global warming and improving agricultural productivity in general.

6. References

Aerni, P., (2009) What is sustainable agriculture? Emperical evidence of diverging views in Switzerland and New Zealand. Ecological Economics 68, 1872-1882.

Allen, J., Thompson, G., (1997) Think global , then think again- Economic globalisation in context. Area 29:213-27.

Baylies, C., (1979) The emergence of indigenous capitalist agriculture: the case of Southern Province, Zambia in Rural-Africana, 4-5, Spring-Fall, pp 65-81.

Benjaminsen, T., Sjaastad, E., (2002) Race for the prize; land transaction and rent appropriation in the Malian Cotton Zone. European Journal of Development Research 14 (2), 129-152.

Benjaminsen, T., Sjaastad, E., (2008) Legal Empowerment for Local Resource Control: Securing Local Resource Rights within Foreign Investment Projects in Africa, London. IIED.

Bertolini, R., (2004) Making information and communication work for food security in Africa. 2020 Africa Conference Brief 11. International Food Policy Research Institute: Washington DC.

Bowe, C., (2007) Potential answers to the adaptation to and mitigation of climate change through the adoption of underutilized crops. Tropical Agriculture Association Newsletter, 27:9-13.

CBD, (2009) Convention on Biological Diversity- CBD. Connecting Biodiversity and Climate Change Mitigation and Adaptation: Report of the Second Ad Hoc Technical Expert Group on Biodiversity and Climate Change. Montreal, Technical Series No. 41, 126 pages.

Clay, D.C., Lewis, L.A., (1990) Land Use, Soil Loss and Sustainable Agriculture in Rwanda. Human Ecology, 18 (2), pp.147-161.

Cotula, L., (2008) Legal Empowerment for Local Resource Control: Securing Local Resource Rights within Foreign Investment Projects in Africa, London. IIED.

Cunguara, B., Darnhofer, I., (2011) Assessing the impact of improved agricultural technologies on household income in rural Mozambique. Food Policy 36.

de Plessis, J., (2005) The growing problem of forced evictions and the crucial importance of community-based, locally appropriate altenatives. Enviroment and Urbanisation 17(1), 123-134.

DFID, (2008) Sustainable Agriculture. Department for International Development (DFID) Research Strategy 2008-2013. Working Paper

Dyke, F., (2008) Conservation Biology: Foundations, Concepts, Applications, 2nd ed. Springer Verlag. pp 478.

ECV, (2011) Exploration, Compensation and Valuation: ADB Policy and International Experience. (http://www.adb.org/Documents/Reports/Capacity-Building-Compensation-Valuation/chap1.pdf) Access 25 August, 2011.

FAO, (1998) Sustaining agricultural biodiversity and agro-ecosystem functions. FAO Headquarters, Rome, Italy.

Fowler, C., (2008) Crop Diversity: Neolithic Foundations for Agriculture's Future Adaptation to Climate Change. Ambio, 498-501.

Gilks, P., (1975) The Dying Lion: Feudalism and Modernization in Ethiopia. Julian Friedmann Publishing Ltd, London.

Guha-Sapir, D., Vos, F., Below, R., Ponserre, S., (2011) Annual Disaster Statistical Review 2010: The Numbers and Trends. Brussels: CRED.

Henao, J., Baanante, C., (2006) Agricultural production and soil nutrient mining in Africa: Implications for resource conservation and policy development. IFDC Technical Bulletin. Muscle Shoals, Ala., U.S.A.: International Fertilizer Development Center.

Higgins, G.M., Kassam, A.H., Naiken, L., Fischer, G., Shah, M.M., (1982) Potential population supporting capacities of lands in the developing world. Food and Agriculture Organization of the United Nations, Rome, Italy. Technical report of project FPA/INT/513, 139 pp

Hulme, M., Kelly, P.M., (1993) Exploring the linkages between climate change and desertification. Environment 35, 4-11

Hutchinson, R., Spooner, B., Walsh, N., (1996) Fighting for Survival: Insecurity, People and the Environment in the Horn of Africa. IUCN (World Conversation Union), Gland, Switzerland.

Isbell, F., (2010) Causes and Consequences of Biodiversity Declines. Nature Education Knowledge 1(11):17.

Jackson, L., Bawa, K., Pascual, U., Perrings, C., (2005) Agrobiodiversity: A new science agenda for biodiversity in support of sustainable agroecosystems. DIVERSITAS Report No. 4. pp.40.

Jansky, L., Chandran, R., (2004) Climate change and sustainable land management: Focus on erosive land degradation. Journal of the World Association of Soil and Water Conservation 4: 17–29.

Jarvis, A., Lane, A., Hijmans, R.J., (2008) The effects of climate change on crop wild relatives. Agriculture Ecosystems & Environment, 126:13-23.

Jones-Walters, L., Mulder, I., (2009) Valuing nature: The economics of biodiversity. Journal for Nature Conservation. 17 (4), 245-247.

Karanja, D.D., Renkow, M., Crawford, E.W., (2003) Welfare effects of maize technologies in marginal and high potential regions of Kenya. Agricultural Economics, 29 (3), 331-341.

Kasanga, K., Kotey, N.A., (2001) Land Management in Ghana. Bulding on Tradition and Modernity. IIED London.

Kassie, M., Shiferaw, B., Muricho, G., (2011) Agricultural Technology, Crop Income and Poverty Alleviation in Uganda. World Development, In Press.

Kijima, Y., Otsuka, K., Sserunkuuma, D., (2008) Assesing the impact of NERICA on income and poverty in central and western Uganda. Agricultural Economics, 38 (3), 327-337.

Kotschi, J., (2007) Agricultural biodiversity is essential for adapting to climate change. Gaia-Ecological Perspectives for Science and Society, 16: 98-101.

KSP, (2005) Kenya Ministry of Agriculture. Ministry of Agriculture Strategic Plan. 2005-2009.

KSRA, (2004) Republic of Kenya. Strategy for Revatilising Agriculture in Kenya (2004-2014). Ministry of Agriculture, Livestock and Fisheries Department.

Levine, A., (2002) Convergence or Convenience? International Conservation NGOs and Development Assistance in Tanzania. World Development 30, 1043-1055.

Lobel, D.B., Burke, M.B., Tebaldi, C., Mastrandrea, M.D., Falcon, W.P., Naylor, R.L., (2008) Prioritizing climate change adaptation needs for food security in 2030. Science, 319: 607-610.

MA, (2005) Millennium Ecosystem Assessment (MA). Ecosystems and Human Well-being: Synthesis. Washington, DC: Island Press.

Minten, B., Barret, C.B., (2008) Agricultural technology, productivity and poverty in Madagascar. World Development, 36 (5), 797-882.

Mvunga, M.P., (1980) The Colonial Foundations of Zambia Land Tenure System. NECZAM, Lusaka.

Myers, N., (1986) The enviromental dimension to security. The Enviromentalist 6 (4): 251-257.

NASA Earth Observatory. NASA Goddard Space Flight Center, USA. http://earthobservatory.nasa.gov/Experiments/PlanetEarthScience/GlobalWarming/GW_InfoCenter_Africa.php Accessed: 21/09/2011

Naeem, S., Bunker, D.E., Hector, A., Loreau, M., Perrings, C., (2009) (Eds), Biodiversity, Ecosystem Functioning, and Human Wellbeing: An Ecological and Economic Perspective. Oxford, UK: Oxford University Press.

Nelson, G.C., Rosegrant, M.W., Koo, J., Robertson, R., Sulser, T., Zhu, T., Ringler, C., Msangi, S., Palazzo, A., Batka, M., Magalhaes, M., Valmonte-Santos, R., Ewing, M., Lee, D., (2009) Climate Change: Impact on Agriculture and Costs of Adaptation. International Food Policy Research Institute (IFPR).

Nigel, R., (1989) Agricultural Extension in Africa; A World Bank Symposium. World Bank, Washington, DC. Descriptive Report 141.

Nkonya, E., Pender, J., Kaizzi, K.C., Edward Kato, Mugarura, S., Ssali, H., Muwonge, J., (2008) Linkages between Land Management, Land Degradation, and Poverty in SSA. The Case of Uganda. International Food Policy Research Institute. Research Report 159.

Peter, P.J., (2011) Climate Change and Security in Africa. World Defense Review. <http://worlddefensereview.com/pham110309.shtml>.

Pintstrup-Anderson, P., Pandya-lorch, R., (1995) Food security and the enviroment. Ecodecision 18:18:22.

Ponte, S., (2001) Policy reforms, market failure and input use in African smallholder agriculture. European Journal of Development Research 13(1):1-29.

Poole, N.D., Kenny, L., (2003) Agricultural market knowledge:System for delivery of a private and public good. Journal of Agricultural Education and Extension, 9. 117.126.

Pritchard, J.M., (1979) Africa: A Study Geography for Advanced Students, Longman Group, Revised Third Edition.

Riverson, J., Gavviriaand, J., Thricutt, S., (1991) Rural Roads in SSA. Lesson learned from World Bank Experience. Technical Paper. No. 141. Washington D.C. World Bank.

Rukuni, M., Blackie, M.J., Eicher, C.K., (1998) Crafting Smallholder-Driven Agricultural Research Systems in Southern Africa. World Development 26, 1073-1087.

Sidahmed, A.E., Yazman, J., (1994) Livestock production and the Environment in lesser developed countries. p. 13-31. In: J. Yazman and A.G. Light (ed.) Proceedings of the International Telecomputer Conference on Perspectives on Livestock Research and Development in Lesser Developed Countries, IDRC, INFORUM, Winrock International, November 1992 - April 1993.

Smaling, E.M.A., Nandwa, S.M., Janssen, B.H., (1997) Soil fertility is at stake. In Replenishing soil fertility in Africa, ed. R. J. Buresh, P. A. Sanchez, and F. Calhoun. SSSA Special Publication 51. Madison, Wisc., U.S.A.: Soil Science Society of America and American Society of Agronomy.

Stiglitz, J., (2003) Globalisation and its discontent. W.W. Norton, New York.

Trapnell, C.G., Clothier, J.N., (1996) The Soils, Vegetation and Traditional Agriculture of Zambia, Volume 1(Central and Western Zambia Ecological Survey 1932-1936); and Volume 2 (by Trapnell, North Eastern Zambia, Ecological Survery 1937-1942).

UN-HABITAT, (2003) United nations human settlements programme (UN-HABITAT). The challenge of slums: global report on human settlements 2003. London: Earthscan.

UN, (2004) United Nations: World Urbanisation Prospects: the 2003 Revision, United Nations Population Division, Department of Economics and Social Affairs, ST/ESA/SER.A/237, New Yorks, 323 pp.

UNCTAD-UNEP, (2008) Organic Agriculture and Food Security in Africa. United Nations Conference on Trade and Development United Nations Environment Programme (UNCTAD-UNEP).

Uphoff, N., (1986) Local Institutional Development: An Analytical Sourcebook, with Cases. Kumarian Press, West Hartford, CN.

Voortman, R.L., Sonneveld, B.G., Keyzer, M.A., (2000) African land ecology: Opportunities and constraints for agricultural development. Center for International Development Working Paper 37. Cambridge, Mass., U.S.A.: Harvard University.

Wake, D.B., Vredenburg, V.T., (2008) "Are we in the midst of the sixth mass extinction? A view from the world of amphibians". Proceedings of the National Academy of Sciences of the United States of America 105: 11466-11473.

WB, (2002) World Bank (WB): Non-governmental organisations and Society.

White, C.M.N., (1959) A Preliminary Survey of Luvale Rural Economy. The Rhodes-Livingstone Papers No 29, Manchester University Press.

Wischmeier, W.H., Smith, D.D., (1978) Predicting Rainfall Erosion Losses, A Guide to Conservation Planning, Agricultural Handbook No 537. USDA, Washington, D.C. pp.1-58.

Wood, A., Stedman-Edward, P., Mang, J., (2000) Root Causes of Biodiversity Loss. Earthscan Publications.

WRI, (1992) World Resources Institute (WRI). World Resources 1992-93: Guide to Global Environment.

Zavale, H., Mabaya, E., Christy, R., (2006) Smallholders' cost efficiency in Mozambique: Implications for improved maize seed adoption. Contributed paper prepared for presentation at the International Association of Agricultural Economists Conference, Gold Coast, Australia, August 12-18.

Part 2

Human Health and Environmental Change Impact

3

Acid Stress Survival Mechanisms of the Cariogenic Bacterium *Streptococcus mutans*

Yoshihisa Yamashita and Yukie Shibata
Section of Preventive and Public Health Dentistry,
Kyushu University Faculty of Dental Science
Japan

1. Introduction

Streptococcus mutans, the major etiological agent in human dental caries, is capable of forming a biofilm, or dental plaque, on the tooth surface (Loesche, 1986; Tanzer et al., 2001). *S. mutans* generates large amounts of acid within dental plaque from fermentable dietary carbohydrates. During meals, the ingestion of carbohydrates causes the pH of the dental plaque to fall below 4.0. Acid accumulation can eventually destroy the crystalline structure of teeth that is the hardest tissue in the human body, leading to the formation of a carious lesion (Quivey et al., 2001). The ability of *S. mutans* to survive in such a severe environment represents one of the most important virulence factors of this microorganism.

The mechanisms of acid tolerance that are most common among Gram-positive bacteria have been proposed to be: i) proton pumps; ii) protection and/or repair of macromolecules; iii) cell-membrane changes; iv) production of alkali; v) regulators; vi) cell density and biofilms; and vii) alteration of metabolic pathways (Fig. 1) (Cotter & Hill, 2003). Many researchers have sought to explain the mechanisms of acid tolerance in *S. mutans*, and various genes contributing to aciduricity in *S. mutans* have been identified. In this chapter, we review those genes that have been reported to be involved in *S. mutans* aciduricity, including those participating in two-component systems and others, especially targeting the *dgk* homolog.

2. Two-component system

Two-component systems (TCSs), prokaryote-specific signal transduction systems, are widespread in prokaryotes and play extensive roles in adaptation to environmental changes. The TCS operon (*tcs*) consists of *hk*, which encodes a sensory histidine kinase (HK), and *rr*, which encodes its cognate response regulator (RR). The HK undergoes autophosphorylation on a histidine residue in response to a specific environmental signal and relays this phosphate group to an aspartic acid residue on the cognate RR. The phosphorylated RR then binds target DNA elements with greater affinity, inducing or repressing the transcription of target genes (Hoch, 2000; Rampersaud et al., 1994). In this way, bacteria are able to adapt to the changes in external environment and to modulate gene expression. TCSs may be responsible for the acid tolerance of *S. mutans*.

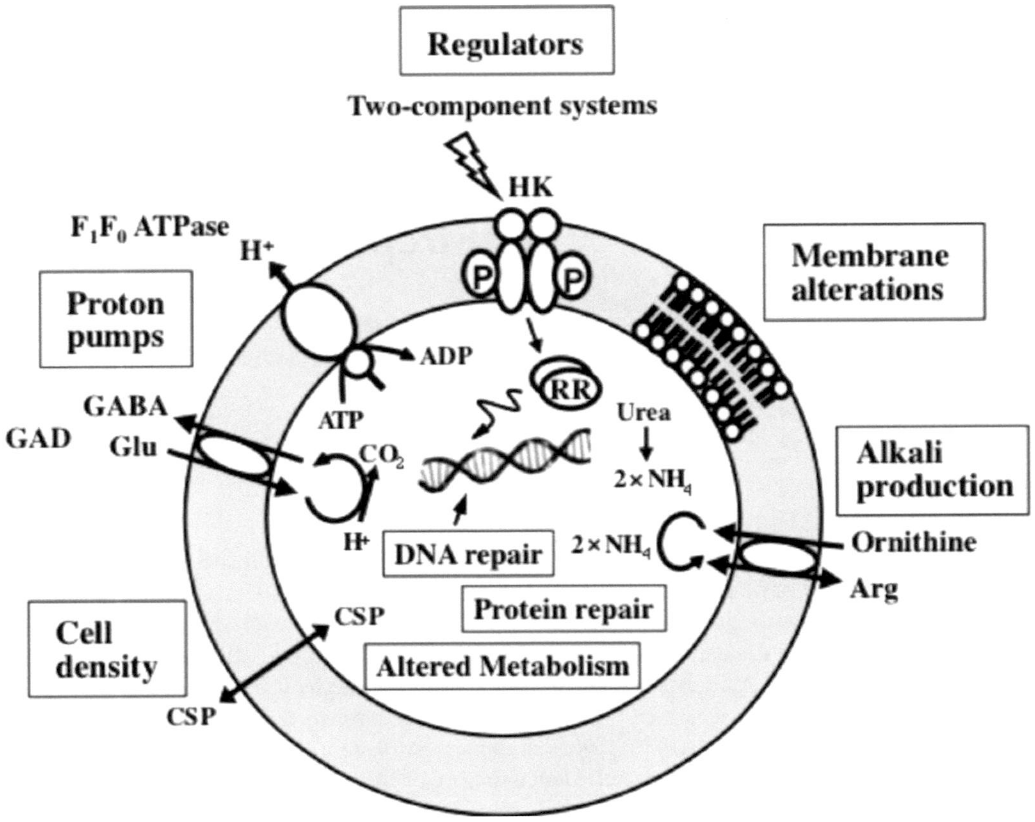

Fig. 1. Acid tolerance mechanisms proposed for gram-positive bacteria. The figure is taken from Cotter & Hill (2003) with some modification.

Analysis of the complete genome sequence of *S. mutans* UA159 suggested the presence of 13 *hk-rr* homologs and one orphan *rr* homolog (Table 1, *smtcs02-15*) (Ajdic et al., 2002). The roles of some specific *tcs* genes in acid tolerance have been evaluated. Li et al. showed that disruption of *smhk13* or *smrr13* resulted in a diminished log-phase acid-tolerance response in *S. mutans* BM71 (Li et al., 2001) and that only *smhk02* of *smtcs02* was involved in the acid-tolerance response of *S. mutans* NG8 (Li et al., 2002). Qi et al. (2004) and Ahn et al. (2006) reported that the *smhk08* mutant exhibited a significant growth defect, whereas the growth of both the *smrr08* mutant and *smhk-rr08* double mutant was similar to that of wild-type UA159 when grown at pH 6.4. Ahn et al. (2006) also showed that all *smhk08*, *smrr08*, or *smhk-rr08* mutants presented growth defects when grown at pH 5.5. Then, Lévesque et al. (2007) systematically inactivated each of the 13 *hk*, but not *rr*, genes in *S. mutans* UA159 and evaluated the roles of the *hk* genes in acid tolerance. They showed that *smhk09* and *smhk14* were involved in *S. mutans* acid tolerance. Furthermore, Biswas et al. (2007) found an additional *tcs* (Table 1, *smtcs01*) in the genome of *S. mutans* UA159 and examined the involvement of 14 *hk* genes in acid tolerance. They showed that only *smhk08* was involved in aciduricity. However, these studies focused only on the role of HKs and so did not provide a comprehensive overview of the role of TCSs in acid tolerance.

tcs genes	*hk gene*, *rr* gene	GenBank Locus Tag [a]	Gene order
smtcs01	*smhk01*	SMU.45	*hk-rr*
	smrr01	SMU.46	
smtcs02	*smhk02*	SMU.486	*hk-rr*
	smrr02	SMU.487	
smtcs03	*smhk03*	SMU.577c	*hk-rr*
	smrr03	SMU.576c	
smtcs04	*smhk04*	SMU.660	*rr-hk*
	smrr04	SMU.659	
smtcs05	*smhk05*	SMU.928	*rr-hk*
	smrr05	SMU.927	
smtcs06	*smhk06*	SMU.1009	*rr-hk*
	smrr06	SMU.1008	
smtcs07	*smhk07*	SMU.1037c	*rr-hk*
	smrr07	SMU.1038c	
smtcs08	*smhk08*	SMU.1128c	*rr-hk*
	smrr08	SMU.1129c	
smtcs09	*smhk09*	SMU.1145c	*rr-hk*
	smrr09	SMU1146c	
smtcs10	*smhk10*	SMU.1516c	*rr-hk*
	smrr10	SMU.1517c	
smtcs11	*smhk11*	SMU.1548c	*hk-rr*
	smrr11	SMU.1547c	
smtcs12	*smhk12*	SMU.1814c	*rr-hk*
	smrr12	SMU.1815c	
smtcs13	*smhk13*	SMU.1916c	*rr-hk*
	smrr13	SMU.1917c	
smtcs14	*smhk14*	SMU.1965c	*hk-rr*
	smrr14	SMU.1964c	
smtcs15	*smrr15*	SMU.1924c	*rr*

tcs, two-component system; *hk*, histidine kinase; *rr*, response regulator.
[a] GenBank locus tag was associated with the *S. mutans* genome at the Oral Pathogen Sequence Database site (http://www.stdgen.lanl.gov/oragen).

Table 1. The *tcs* genes identified in the *S. mutans* UA159 genome.

Therefore, we systematically constructed *rr* deletion mutants and *hk-rr* double mutants of *S. mutans* UA159 and examined the effect on acid tolerance (Kawada-Matsuo et al., 2009). Thirteen *rr* mutants and twelve *hk-rr* double mutants were obtained, the exceptions being *smrr10*, *smtcs10*, *smrr12*, and *smtcs12*. The derivation of null mutations of these genes was unsuccessful, probably due to a loss of viability of these mutants. To examine the effects of these *rr* mutations on the acid tolerance of *S. mutans*, wild-type UA159 and the 25 mutants were grown in brain–heart infusion (BHI) broth adjusted to pH 7.2 or pH 5.5. Growth curves were generated, and the mid-log-phase doubling time was determined. All *rr* and *hk-rr* mutants grew similarly to wild-type UA159 at pH 7.2. However, as shown in Table 2, deletion of four *rr* genes (*smrr03*, *smrr05*, *smrr08*, and *smrr13*) caused significantly decreased growth rates compared with that of wild-type UA159 when grown at pH 5.5. The growth rates of the *hk-rr* double mutants were similar to those of the corresponding *rr* mutants, and the differences in doubling time between them were not significant. The finding that *smrr08* and *smrr13* were involved in the acid tolerance of *S. mutans* is consistent with previous findings (Ahn et al., 2006; Li et al., 2001). On the other hand, *smrr03* and *smrr05* were, for the first time, demonstrated to be involved in *S. mutans* UA159 acid tolerance. However, Lévesque et al. (2007) and Biswas et al. (2007) showed that inactivation of their cognate *hk* genes did not affect acid tolerance. To confirm whether only the *rr* gene is involved in acid tolerance in the *smtcs03* and *smtcs05* mutants, the *hk* genes of *smtcs03* and *smtcs05* were individually inactivated, and the acid tolerance ability of these mutants compared. As shown in Table 3, the *smhk03* mutant exhibited a decreased growth rate compared with the wild type when grown at pH 5.5. This was not consistent with previous results. In contrast, the *smhk05* mutant grew similarly to wild-type UA159 at pH 5.5, as shown in previous studies, whereas the *smrr05* mutant and *smhk-rr05* double mutant exhibited reduced growth rates.

	Doublimg time (min) [a] in:	
UA159 [b]	123.8 ± 9.5	
tcs genes	$hk^{+}rr^{-}$	$hk^{-}rr^{-}$
smtcs01	116.8 ± 10.0	117.7 ± 10.3
smtcs02	131.5 ± 8.8	132.6 ± 5.9
smtcs03	146.4 ± 7.5*	148.5 ± 8.3*
smtcs04	127.4 ± 7.6	122.5 ± 3.7
smtcs05	160.1 ± 11.6**	155.7 ± 7.8*
smtcs06	132.8 ± 3.2	125.1 ± 10.2
smtcs07	131.0 ± 3.2	122.3 ± 7.2
smtcs08	146.3 ± 11.9*	145.2 ± 7.7*
smtcs09	126.3 ± 5.2	129.9 ± 5.0
smtcs11	126.4 ± 7.1	131.7 ± 11.0
smtcs13	141.5 ± 8.8*	140.9 ± 2.6*
smtcs14	132.5 ± 3.7	131.7 ± 12.1
smtcs15 [c]	127.6 ± 7.1	–

[a]Doubling time (Td) was calculated based on the formulas ln Z – ln Z0 = k (t – t0), where k is the growth rate, and g = 0.693/k, where g is the doubling time. Values are the mean ± standard deviation obtained from three independent experiments.
[b]Wild-type strain
[c]Orphan *rr*
* Significant increase from Td of wild-type UA159 by Tukey's HSD, $p < 0.05$
** Significant increase from Td of wild-type UA159 by Tukey's HSD, $p < 0.01$

Table 2. Doubling times of *rr* or *hk-rr* deletion mutants at pH 5.5.

Generally, a signal sensed by a HK is thought to be transmitted to the cognate RR via transfer of phosphoryl groups, and deletion of either the *hk* or *rr* should generate a similar phenotype. However, we found that only the *rr* of *smtcs05* was involved in *S. mutans* acid tolerance. Furthermore, as mentioned above, Li et al. (2002) reported that the *hk*, but not the *rr*, of *smtcs02* was involved in the acid tolerance of *S. mutans* NG8. Qi et al. (2004) and Ahn et al. (2006) also reported that the *smhk08* mutant showed a significant growth defect, whereas the growth of both the *smrr08* mutant and *smhk-rr08* double mutant was similar to that of wild-type UA159. These results suggested involvement of several TCSs in *S. mutans* aciduricity via cross-talk between different TCS components. After all, among 15 TCSs, seven (Smtcs02, 03, 05, 08, 09, 13, and 14) appear to be involved in *S. mutans* aciduricity. Nevertheless, inactivation of no single TCS caused a complete loss of acid tolerance. Therefore, other TCSs that are definitively related to *S. mutans* acid tolerance may exist, but they cannot be identified by homology searching.

tcs genes	Strain	*hk*/*rr*	Doubling time (min) in:		
			BHI pH 7.2	BHI pH 5.5	
smtcs03	SMHK03	-/+	53.6 ± 4.1	153.4 ± 7.9	
	SMRR03	+/-	50.0 ± 2.7	146.4 ± 7.5	
	SMTCS03	-/-	57.1 ± 8.6	148.5 ± 8.3	
smtcs05	SMHK05	-/+	51.0 ± 3.4	128.8 ± 12.9	*
	SMRR05	+/-	49.7 ± 2.5	160.1 ± 11.6	*
	SMTCS05	-/-	51.5 ± 4.6	155.7 ± 7.8	

*$p < 0.05$ (Tukey's HSD)

Table 3. Doubling times of *hk*, *rr*, and *hk-rr* deletion mutants of *smtcs03* and *smtcs05* at pH 7.2 and pH 5.5.

3. Genes other than TCS involved in *S. mutans* acid tolerance

Many studies have implicated genes other than those involved in TCS in *S. mutans* aciduricity. Table 4 summarizes the characteristics of 14 such genes, of which the functions of the products of only nine have been experimentally verified.

These are *aguA*, encoding an agmatine deiminase that is involved in alkali production (Griswold et al., 2004); *dltC*, involved in the synthesis of D-alanyl-lipoteichoic acid, which is associated with alteration of membrane composition (Boyd et al., 2000); *gluA*, encoding a glucose-l-phosphate uridylyltransferase involved in the synthesis of UDP-D-glucose (Yamashita et al., 1998); *lgl*, encoding a lactoylglutathione lyase involved in the detoxification of methylglyoxal (Korithoski et al., 2007); *luxS*, encoding a S-ribosylhomocysteine lyase involved in autoinducer AI2 synthesis (Wen and Burne, 2004); and *uvrA*, encoding an excinuclease ABC subunit A that is involved in DNA repair (Hanna et al., 2001). Mutation of three genes (*ffh*, *ftsY*, *yidC2* genes) that are involved in the signal recognition particle pathway, significantly reduced H^+/ATPase specific activity compared with that of the wild type (Hasona et al., 2005). However, how these functions contribute to *S. mutans* aciduricity remains unclear. Furthermore, the functions of the five remaining gene products were predicted based on DNA sequence homology, and so estimation of their role in *S. mutans* aciduricity is much more difficult.

Gene	GenBank Locus Tag	Function	Evidence	Determination of the response against low pH	Reference
aguA	SMU.264	Agmatine deiminase	Testified experimentally	Not determined	Griswold et al., 2004
brpA	SMU.410	Transcriptional regulator	Putative	Acid tolerance & Acid killing	Wen et al., 2006
clpP	SMU.1672c	Clp protease (Serine protease)	Putative	Acid tolerance	Lemos & Burne, 2002
dltC	SMU.1689c	D-alanyl carrier protein	Testified experimentally	Acid tolerance & Acid killing & Acid tolerance response	Boyd et al., 2000
ffh	SMU.1060c	Signal recognition particle	Testified experimentally (partially)	Acid tolerance	Kremer et al., 2001
ftsY	SMU.744	Signal recognition particle receptor	Testified experimentally (partially)	Acid tolerance	Hasona et al., 2005
glrA	SMU.1035	ABC transporter ATP-binding-protein	Putative	Acid tolerance	Cvitkovitch et al., 2000
gluA	SMU.322c	Glucose-l-phosphate uridylyltransferase	Testified experimentally	Acid tolerance	Yamashita et al., 1998
htrA	SMU.2164	Serine protease	Putative	Acid tolerance (agar plate)	Biswas & Biswas, 2005
lgl	SMU.1603	Lactoylglutathione lyase	Testified experimentally	Acid tolerance & Acid tolerance response	Korithoski et al., 2007
luxS	SMU.474c	S-ribosylhomo-cysteine lyase	Testified experimentally	Acid tolerance & Acid killing & Acid tolerance response	Wen & Burne, 2004
ropA	SMU.91	Peptidyl-prolyl isomerase, trigger factor	Putative	Acid killing	Wen et al., 2005
uvrA	SMU.1851c	UV repair excinuclease	Testified experimentally	Acid tolerance & Acid tolerance response	Hanna et al., 2001
yidC2	SMU.1727	Oxa1(or A)-like protein	Testified experimentally (partially)	Acid tolerance	Dong et al., 2008

Table 4. Genes reported to be involved in *S. mutans* acid tolerance.

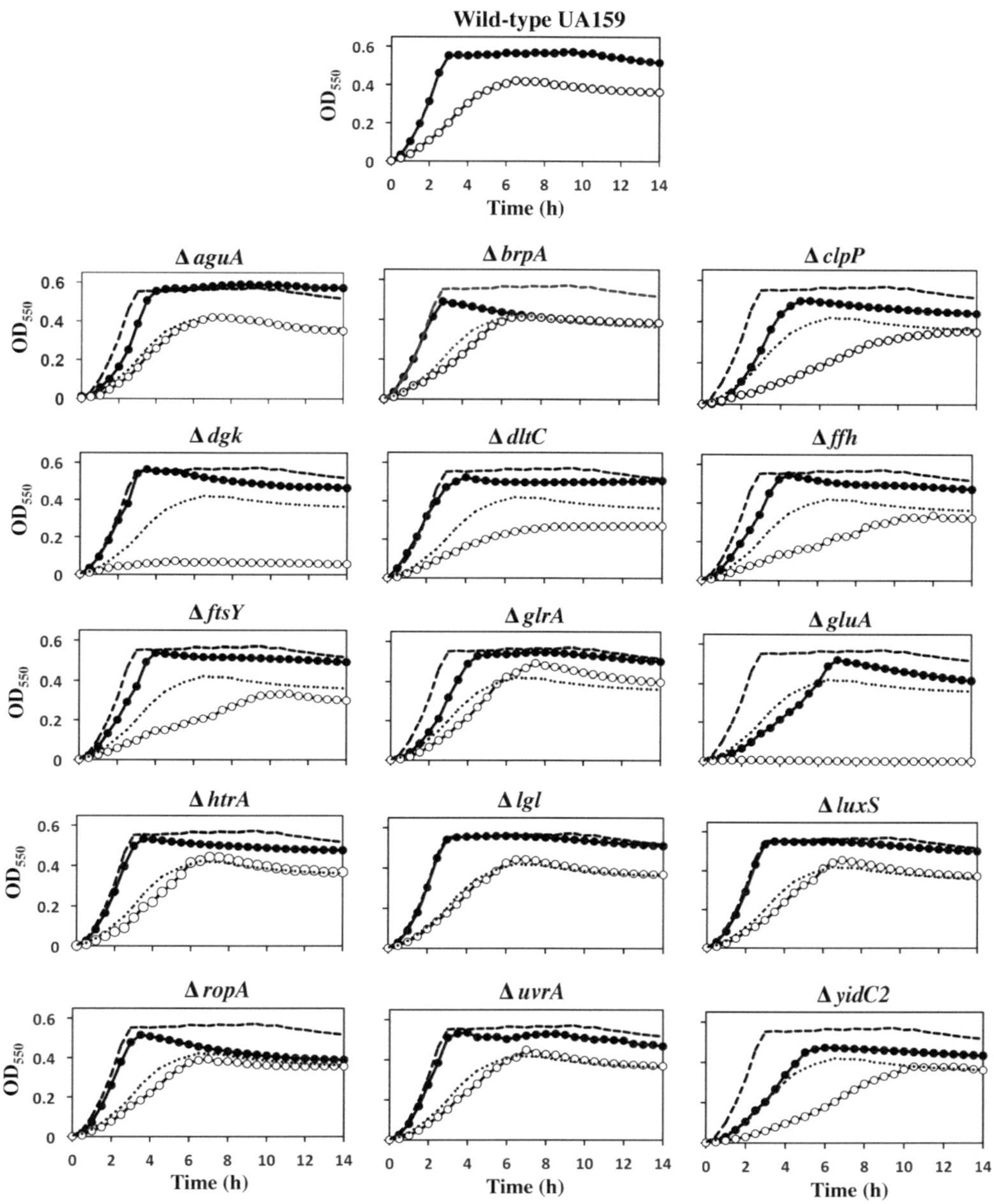

Fig. 2. Growth curves of *S. mutans* UA159 and mutant strains grown in BHI medium at pH 7.4 (●) or pH 5.5 (○). Growth was defined as the increase in OD_{550}, and was calculated by subtraction of OD_{550} at the initiation of growth from that at the times indicated. Data represent the means of three independent experiments. Graphs of mutant strains also represent the growth curves of wild-type UA159 at pH 7.4 (— — —) and pH 5.5 (·····) as controls.

To examine the extent to which a gene contributes to *S. mutans* acid tolerance, 14 mutants in which one of the genes listed in Table 4 was inactivated and the *dgk* mutant were constructed from *S. mutans* UA159, and their growth at pH 7.4 and at pH 5.5 was compared with that of the wild type (Fig. 2) (Shibata et al., 2011). Inactivation of *aguA, brpA, glrA, htrA, lgl, luxS, ropA,* or *uvrA* did not significantly affect the acid tolerance of *S. mutans* as compared with wild-type UA159 when grown at pH 5.5. Three types of method of assessing the acid tolerance of *S. mutans* are available. Simple acid tolerance is evaluated by a procedure in which over-night cultures or log-phase cells grown at neutral pH are subcultured in media or on plates at neutral and acidic pH. Another method is acid killing, in which log-phase cells grown at neutral pH are incubated at a lethally acidic pH, and then viability is determined by plate counts. The final method is determination of the acid tolerance response in which viability is estimated by plating after incubation at neutral or acidic pH of log-phase cells grown at neutral pH followed by incubation at killing pH. We used the first method, and so any discrepancy between this and other studies may derive from differences in the method used. However, comparing all of the mutants in terms of the most basic characteristics of acid tolerance is a worthwhile endeavor. Of course, a comparison using the other criteria is important, and will be performed as part of our next effort.

Notably, the *dgk* and the *gluA* mutants grew extremely slowly at pH 5.5, although the *clpP, dltC, ffh, ftsY,* and *yidC2* mutants also displayed significant reductions in growth rate at pH 5.5 compared with the wild-type UA159. However, only the *brpA, dgk, dltC, htrA, lgl, luxS, ropA,* and *uvrA* mutants showed growth rates comparable to the wild-type strain at pH 7.4. These findings suggest that the reduction in growth rates of the *clpP, ffh, ftsY, yidC2,* and *gluA* mutants at acidic pH values might be derived from reduced viability and not specifically related to acid tolerance. Therefore, *dltC* and *dgk* are likely specifically involved in acid tolerance. Of these two genes, the striking reduction in growth rate of the *dgk* mutant at acidic pH indicates that *dgk* is of great interest for elucidating the acid-tolerance mechanism of *S. mutans.*

4. Diacylglycerol kinase

Diacylglycerol kinase (Dgk) catalyses the ATP-dependent phosphorylation of *sn*-1,2-diacylglycerols, resulting in production of phosphatidic acid. In eukaryotic cells, diacylglycerol and phosphatidic acid are immediate second cellular messengers responding to extracellular signals, suggesting that Dgk is a key enzyme in cellular signal transduction (Moolenaar et al., 1986; Murayama & Ui, 1987; Nishizuka, 1984; Topham & Prescott, 1999). Among bacterial Dgk, only that of *Escherichia coli* has been well-characterized, and it is a small integral membrane protein with a molecular mass of 13.2 kDa. This enzyme functions in the recycling of the diacylglycerol produced during turnover of membrane phospholipids (Hasin & Kennedy, 1982; Rotering & Raetz, 1983) and plays an important physiological role in responding to environmental stress as well as its role in eukaryotic cells (Raetz & Newman, 1979; Walsh et al., 1986). On the other hand, the *S. mutans* Dgk homolog has a molecular mass of 15.3 kDa and comprises 137 amino acids. It is interesting that insertion of the transposon Tn*916* into the codon for the tenth amino acid from the C terminus of the Dgk homolog resulted in defective growth of the mutant (GS5Tn1) at acidic pH values

(Yamashita et al., 1993). In addition to attenuation of aciduricity, this mutant possessed reduced resistance to high osmolarity and temperature (Yamashita et al., 1993). The C terminus of the Dgk homolog may thus play an important role in signal transduction during environment stress.

To evaluate how the C terminus of Dgk contributes to *S. mutans* acid tolerance, we sequentially truncated amino acids from the C terminus of Dgk and finally constructed 11 mutants termed UADGK0–10, expressing Dgk0–10 (Fig. 3) (Shibata et al., 2009). The mutants showed no significant difference in growth rate at neutral pH (doubling times: 53.8 to 61.6 min; Table 5). Most, with the exception of UADGK0 to UADGK2, showed a reduction in growth rate at pH 5.5 compared with the wild type (Table 5 and Fig. 4). UADGK3, in which three amino acid residues had been deleted from the C-terminus of Dgk, showed a slight reduction in growth rate. Subsequent deletion of amino acids from the C-terminus resulted in further reductions in growth rate at acidic pH. Indeed UADGK4, UADGK5, and UADGK6 had significantly increased doubling times ($p < 0.05$, $p < 0.001$, and $p < 0.0001$, respectively) compared with UADGK0. UADGK7, in which seven amino acid residues had been deleted, showed extremely limited growth in the first 9 hours. Further truncation of the C-terminus of Dgk (UADGK8 to UADGK10) resulted in no growth at pH 5.5. These results suggest that the C-terminal of the Dgk homolog is indispensable for its function in aciduricity of *S. mutans*. We further constructed two additional UA159 *dgk* mutants, UADGK11 and UADGK12 (Fig. 3) to evaluate the function of truncated Dgk. There were only negligible differences in the growth rates of these two mutants at pH 5.5, 5.8, or 6.3, compared with that of UADGK10.

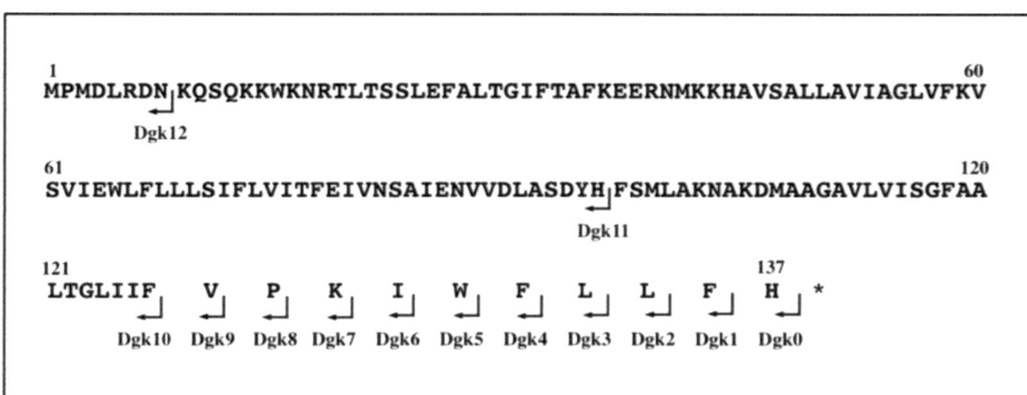

Fig. 3. Representation of the truncated Dgk proteins used. The deduced amino acid sequence of the *dgk* gene from *S. mutans* UA159 is presented. The terminal amino acid of the truncated Dgk expressed in each UA159 *dgk* mutant and each *E. coli* RZ transformant is indicated along the sequences by a curved arrow, which indicates that the sequence is deleted from the right up to this site. All the truncated Dgk proteins names were changed from those of previous paper (Shibata et al., 2009) to help readers understand.

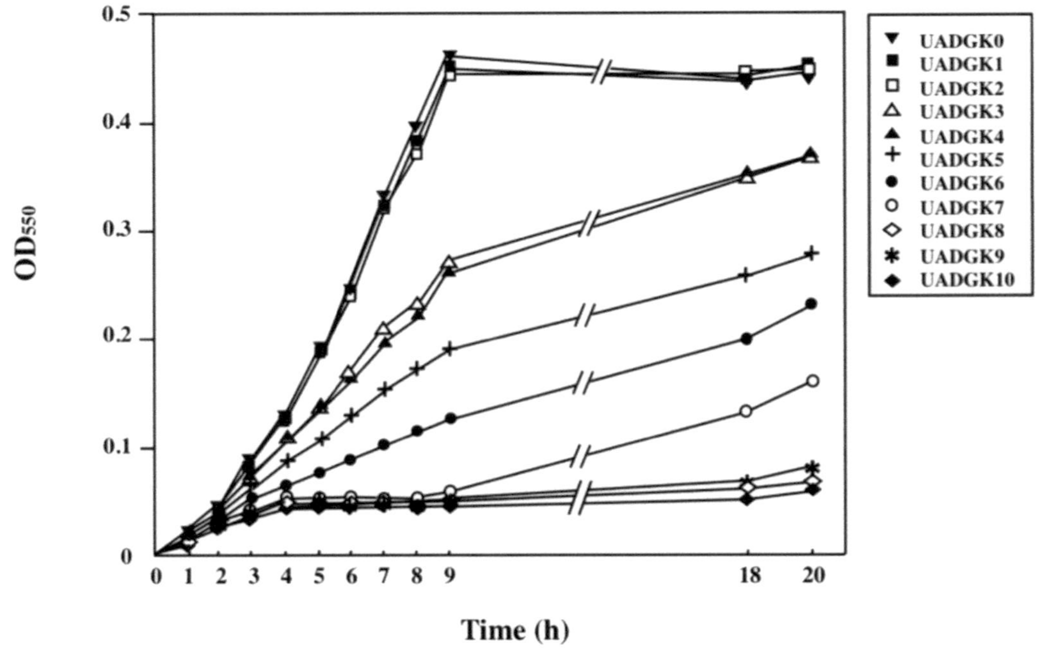

Fig. 4. Growth curves of *S. mutans* UA159 *dgk* mutants grown in BHI medium at pH 5.5. Growth was defined as the increase in OD_{550}, and was calculated by subtraction of OD_{550} at the initiation of growth from that at the times indicated. Data represent the means of three independent experiments. All the mutants names were changed from those of previous paper (Shibata et al., 2009) to help readers understand.

Strain	Doubling time (min) in:	
	BHI pH 7.45	BHI pH 5.5
UADGK0	57.8 ± 4.4	132.3 ± 12.9
UADGK1	58.2 ± 6.4	131.2 ± 11.7
UADGK2	61.2 ± 5.8	133.6 ± 11.7
UADGK3	61.6 ± 1.8	177.6 ± 19.0
UADGK4	55.5 ± 3.4	202.5 ± 24.4*
UADGK5	57.5 ± 2.5	248.9 ± 32.2**
UADGK6	55.3 ± 5.4	271.3 ± 34.6 ***
UADGK7	55.1 ± 6.0	> 1000 [a]
UADGK8	53.8 ± 3.7	> 1000 [a]
UADGK9	58.4 ± 6.7	> 1000 [a]
UADGK10	57.0 ± 6.1	> 1000 [a]

Differences in the doubling time between UADGK0 and UADGK1-10 were analyzed by Bonferroni multiple comparison test (*, $p < 0.05$; **, $p < 0.001$; ***, $p < 0.0001$).
[a]Statistical analyses were not carried out because of too slow growth rates.
All the mutants names were changed from those of previous paper (Shibata et al., 2009) to help readers understand.

Table 5. Effect of low pH on growth of *S. mutans* UA159 *dgk* mutants.

We next constructed recombinant Dgk proteins corresponding to *S. mutans* strains UADGK10, UADGK7, UADGK5, and UADGK0 utilizing *E. coli* strains RZDGK10, RZDGK7, RZDGK5, and RZDGK0, respectively. The kinase activity in cell lysates of *E. coli* transformants was examined by an octyl glucoside mixed-micelle assay (Preiss et al., 1986), using undecaprenol as a substrate because of it has a higher substrate specificity for the *S. mutans* Dgk homolog compared with diacylglycerol (Lis & Kuramitsu, 2003).

As shown in Fig. 5A, whereas the full-size *S. mutans* Dgk protein expressed in RZDGK0 catalyzed a high level of phosphorylation of undecaprenol, the Dgk missing five amino acid residues from the C terminus expressed in RZDGK5 exhibited markedly reduced kinase activity. Furthermore, RZDGK7 (seven amino acids missing from the C terminus) exhibited much weaker kinase activity than did RZDGK5. The deletion of 10 C-terminal amino acid residues of Dgk in RZDGK10 resulted in a total lack of kinase activity. These differences were confirmed by quantitative analysis (Fig. 5B). These data indicate that the C-terminus of the *S. mutans* Dgk homolog plays an important role in kinase activity and may harbor residues required for catalysis. Alternatively, incorrect folding of the protein due to the missing C-terminal residues may cause loss of kinase activity. Therefore, its catalysis of undecaprenol phosphorylation is closely related to *S. mutans* acid tolerance.

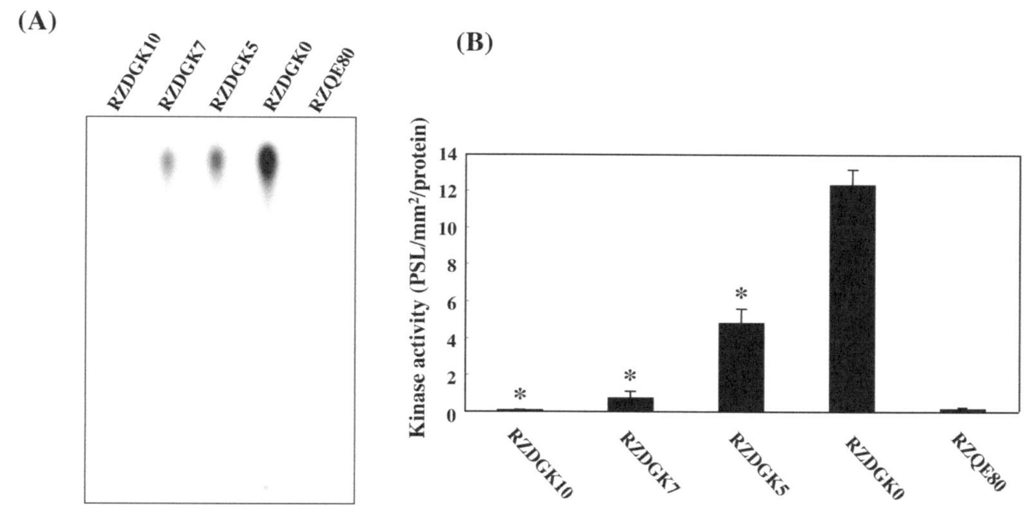

Fig. 5. Effect of deletion of the C-terminal tail of Dgk on undecaprenol kinase activity. (A) Comparison of undecaprenol kinase activity of the full-size Dgk and various C-terminally truncated forms of Dgk. The undecaprenol kinase activity in the lysates from *E. coli* RZ cells was determined using an octyl glucoside mixed-micelle assay. (B) Quantification analysis of the kinase assay. Quantification was carried out by normalization of radioactive bands in the kinase assay using the protein level. Vertical bars represent standard deviation. Differences in kinase activity between RZDGK0 and RZDGK5, RZDGK7, or RZDGK10 were analyzed by Student's *t* test (*, $p < 0.0001$). All the *E.coli* strains names were changed from those of previous paper (Shibata et al., 2009) to help readers understand.

Moreover, the importance of the C-terminal end of Dgk in *S. mutans* acid tolerance was examined in a specific pathogen-free animal model (Table 6). The *dgk* mutant strain clearly displayed a significant reduction in smooth-surface caries compared with the wild type ($p < 0.005$). In contrast, no significant difference in plaque extent was observed between the wild-type and *dgk* mutant strains. These results suggest that aciduricity regulated by the *dgk* gene product might play a critical role in *S. mutans* virulence.

Treatment	Plaque extent (Δ)	Initial dentinal fissures (ΔΔ)	Advanced dentinal fissures (ΔΔ)	Smooth-surface caries (ΔΔΔ)	Total bacteria CFU (10^7)	Total streptococci CFU (10^7)	Total *S. mutans* CFU (10^7)
Water control	2.8 ± 0.63[a b]	9.5 ± 1.72[a]	6.6 ± 2.80[a b]	0.5 ± 0.97[a]	4.4 ± 2.36[a]	2.7 ± 2.26[a]	ND[a b]
UA159 (wt)	1.1 ± 0.32[a]	11.5 ± 1.27[a]	10.8 ± 1.62[a]	9.5 ± 6.55[a c]	7.9 ± 3.21[a]	7.4 ± 2.79[a c]	4.0 ± 2.11[a c]
UADGK10 (*dgk*)	1.5 ± 0.53[b]	10.6 ± 0.84	9.3 ± 1.2[b]	2.5 ± 2.68[c]	6.4 ± 2.40	4.4 ± 2.53[c]	2.3 ± 1.20[b c]

ND, Not determined. Δ, 4 units at risk; ΔΔ, 12 fissures at risk; ΔΔΔ, 20 units at risk.
[a]Significant difference between water control and UA159 (wt), $p < 0.05$.
[b]Significant difference between UADGK1 (*dgk*) and water control, $p < 0.05$.
[c]Significant difference between UA159 (wt) and UADGK1 (*dgk*), $p < 0.05$.
The mutant name was changed from those of previous paper (Shibata et al., 2009) to help readers understand.

Table 6. Influence of *dgk* deletion on smooth-surface plaque extent, initial and advanced dentinal fissure lesions, smooth-surface caries, and colonization properties.

5. A potential target for anti-caries chemotherapy

Development of an effective anti-caries agent is the ultimate goal of our work. Considering the characteristics of known mutants, the Dgk homolog seems to be the most promising target for anti-caries agents. Dgks have been extensively studied in mammals, and several inhibitory compounds, e.g., R59022 and R59949 (Fig. 6), have been reported. In contrast, inhibitors of prokaryotic Dgk have not yet been elucidated.

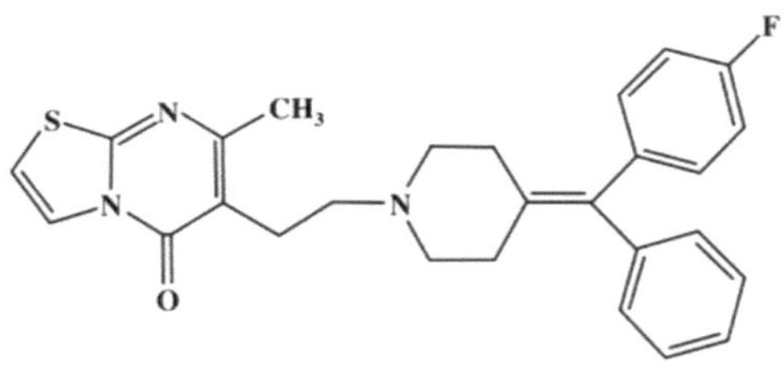

R59022

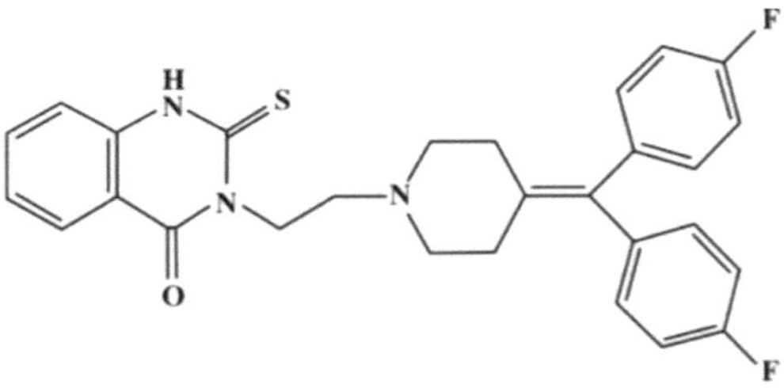

R59949

Fig. 6. Structures of R59022 and R59949.

When first attempting to discover inhibitors of prokaryotic Dgk, we tested the effects of R59022 and R59949 on the growth of *S. mutans* (Shibata et al., 2011). Although neither R59022 nor R59949 influenced growth at pH 7.4, R59949, but not R59022, showed a significant inhibitory effect at acidic pH (Fig. 7). Inhibition by R59949 increased by 13, 29, 58,

68, and 78% at pH 5.4, 5.3, 5.2, 5.1, and 5.0, respectively (Fig. 7A). These findings were particularly interesting because R59022 and R59949 were used at concentrations of 100 µM and 25 µM, respectively, due to the limited solubility of R59949.

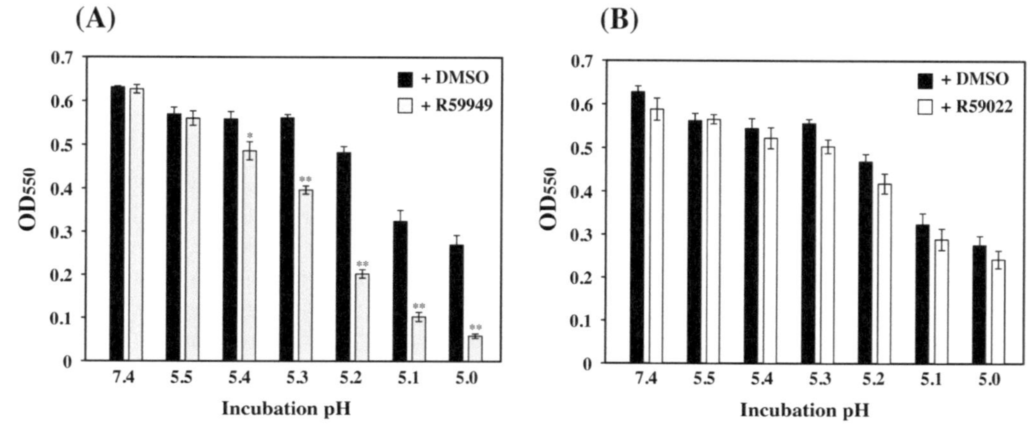

Fig. 7. Effect of R59949 (A) and R59022 (B) on the growth of *S. mutans*. Data represent the mean ± standard deviation. Differences in growth rate between cells cultured in the presence and absence of Dgk inhibitor were analyzed using Student's *t* test. *, $p < 0.05$; **, $p < 0.0001$.

Furthermore, we examined the inhibitory effects of R59022 and R59949 on the kinase activity of *S. mutans* Dgk. Neither R59022 nor R59949 inhibited kinase activity at pH 7.4; this is in agreement with their lack of effect on *S. mutans* growth at neutral pH. As mentioned above, R59949 significantly inhibited the growth of *S. mutans* at acidic pH values (below 5.4). When evaluating the effect of R59949 on enzyme activity at acidic pH, it is important to know the intracellular pH of *S. mutans* cells; the intracellular pH of *S. mutans* cells was 6.4 when cultured in broth at pH 5.2. Therefore, we determined the inhibitory effect of R59949 and R59022 on *S. mutans* Dgk kinase activity at pH 6.4. R59949, but not R59022, inhibited kinase activity with undecaprenol as a substrate by around 20% (Fig. 8).

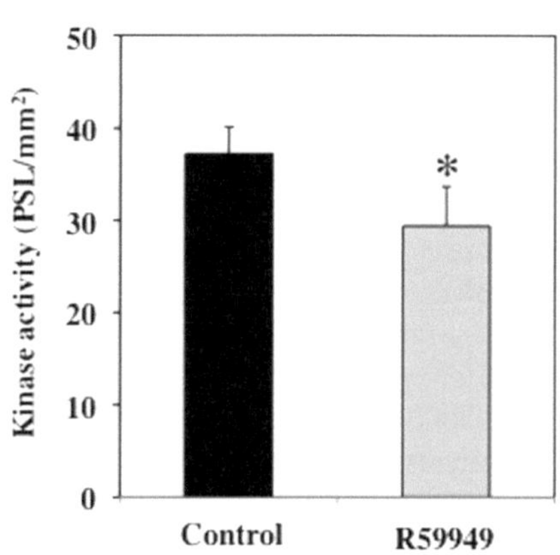

Fig. 8. Effect of R59949 on kinase activity with undecaprenol as a substrate. Data represent the mean ± standard deviation. Differences in kinase activity between cells cultured in the presence and absence of R59949 were analyzed by Student's *t* test. *, $p < 0.05$.

S. mutans Dgk is inherently different from mammalian Dgk in terms of its molecular size, molecular structure, and substrate specificity. However, it is interesting that R59949 inhibits the enzymatic activity of *S. mutans* Dgk even with undecaprenol as the substrate. Additionally, the difference in inhibitory activity between R59949 and R59022 means that a comparison of their molecular structure may lead to discovery of further potent Dgk inhibitors specific for prokaryotic enzymes, that is, new anti-caries agents.

6. Conclusion

The reduction of environmental pH in dental plaque by the cariogenic microorganisms is important step in the development of dental caries. The cariogenic microorganisms should survive in a relentless environment produced by themselves in order to exhibit their maximum virulence. In this chapter, we described the acid tolerance characteristics of the cariogenic microorganism, *S. mutans*. TCSs seem to be the most suitable system for adaptation to environmental conditions. However, no TCS seems to be definitively responsible for *S. mutans* acid tolerance. At present, identification of TCS depends on gene homology searching, which may not identify all genes encoding TCS that contribute to *S. mutans* acid tolerance.

We focus on *dgk* because it is the most promising contributor to *S. mutans* acid tolerance when assessed using a simple acid tolerance assay. Although the precise mechanism by which the gene product is involved in acid tolerance has not yet been elucidated, *dgk* is the only gene whose product has been definitively implicated in cariogenicity in an animal model. Furthermore, potential specific inhibitors of the gene product have been introduced. This fact may aid in development of next-generation anti-caries therapies based on the ability of this microorganism to adapt to environmental conditions.

However, much detail of the acid tolerance mechanisms of *S. mutans* remains unknown, and so further study is required.

7. References

Ahn, S.J., Wen, Z.T., & Burne, R.A. (2006). Multilevel control of competence development and stress tolerance in *Streptococcus mutans* UA159. *Infect Immun*, 74, 3, 1631-1642

Ajdic, D., McShan, W.M., McLaughlin, R.E., Savic, G., Chang, J., Carson, M.B., Primeaux, C., Tian, R., Kenton, S., Jia, H., et al. (2002). Genome sequence of *Streptococcus mutans* UA159, a cariogenic dental pathogen. *Proc Natl Acad Sci U S A*, 99, 22, 14434-14439

Biswas, I., Drake, L., Erkina, D., & Biswas, S. (2007). Involvement of sensor kinases in the stress tolerance response of *Streptococcus mutans*. *J Bacteriol*, 190, 1, 68-77

Biswas, S. & Biswas, I. (2005). Role of HtrA in surface protein expression and biofilm formation by *Streptococcus mutans*. *Infect Immun*, 73, 10, 6923-6934

Boyd, D.A., Cvitkovitch, D.G., Bleiweis, A.S., Kiriukhin, M.Y., Debabov, D.V., Neuhaus, F.C., & Hamilton, I.R. (2000). Defects in D-alanyl-lipoteichoic acid synthesis in *Streptococcus mutans* results in acid sensitivity. *J Bacteriol*, 182, 21, 6055-6065

Cvitkovitch, D.G., Gutierrez, J.A., Behari, J., Youngman, P.J., Wetz, J.E., Crowley, P.J., Hillman, J.D., Brady, L.J., & Bleiweis, A.S. (2000). Tn*917-lac* mutagenesis of *Streptococcus mutans* to identify environmentally regulated genes. *FEMS Microbiol Lett*, 182, 1, 149-154

Cotter, P.D. & Hill, C. (2003). Surviving the acid test: responses of gram-positive bacteria to low pH. *Microbiol Mol Biol Rev*, 67, 3, 429-453

Dong, Y., Palmer, S.R., Hasona, A., Nagamori, S., Kaback, H.R., Dalbey, R.E. & Brady, L.J. (2008). Functional overlap but lack of complete cross-complementation of *Streptococcus mutans* and *Escherichia coli* YidC orthologs. *J Bacteriol*, 190, 7, 2458-2469

Griswold, A.R., Chen, Y.Y., & Burne, R.A. (2004). Analysis of an agmatine deiminase gene cluster in *Streptococcus mutans* UA159. *J Bacteriol*, 186, 6, 1902-1904

Hanna, M.N., Ferguson, R.J., Li, Y.H., & Cvitkovitch, D.G. (2001). *uvrA* is an acid-inducible gene involved in the adaptive response to low pH in *Streptococcus mutans*. *J Bacteriol*, 183, 20, 5964-5973

Hasin, M. & Kennedy, E.P. (1982). Role of phosphatidylethanolamine in the biosynthesis of pyrophosphoethanolamine residues in the lipopolysaccharide of *Escherichia coli*. *J Biol Chem*, 257, 21, 12475- 12477

Hasona, A., Crowley, P.J., Levesque, C.M., Mair, R.W., Cvitkovitch, D.G., Bleiweis, A.S., & Brady, L.J. (2005). Streptococcal viability and diminished stress tolerance in mutants lacking the signal recognition particle pathway or YidC2. *Proc Natl Acad Sci U S A*, 102, 48, 17466-17471

Hoch, J.A. (2000). Two-component and phosphorelay signal transduction. *Curr Opin Microbiol*, 3, 2, 165-170

Kawada-Matsuo, M., Shibata, Y., & Yamashita, Y. (2009). Role of two component signaling response regulators in acid tolerance of *Streptococcus mutans*. *Oral Microbiol Immunol*, 24, 2, 173-176

Korithoski, B., Levesque, C.M., & Cvitkovitch, D.G. (2007). Involvement of the detoxifying enzyme lactoylglutathione lyase in *Streptococcus mutans* aciduricity. *J Bacteriol*, 189, 21, 7586-7592

Kremer, B.H., van der Kraan, M., Crowley, P.J., Hamilton, I.R., Brady, L.J. & Bleiweis, A.S. (2001). Characterization of the *sat* operon in *Streptococcus mutans*: evidence for a role of Ffh in acid tolerance. *J Bacteriol*, 183, 8, 2543-2552

Lemos, J.A. & Burne, R.A. (2002). Regulation and Physiological Significance of ClpC and ClpP in *Streptococcus mutans*. *J Bacteriol*, 184, 22, 6357-6366

Lévesque, C.M., Mair, R.W., Perry, J.A., Lau, P.C., Li, Y.H., & Cvitkovitch, D.G. (2007). Systemic inactivation and phenotypic characterization of two-component systems in expression of *Streptococcus mutans* virulence properties. *Lett Appl Microbiol*, 45, 4, 398-404

Li, Y.H., Hanna, M.N., Svensater, G., Ellen, R.P., & Cvitkovitch, D.G. (2001). Cell density modulates acid adaptation in *Streptococcus mutans*: implications for survival in biofilms. *J Bacteriol*, 183, 23, 6875-6884

Li, Y.H., Lau, P.C., Tang, N., Svensater, G., Ellen, R.P., & Cvitkovitch, D.G. (2002). Novel two-component regulatory system involved in biofilm formation and acid resistance in *Streptococcus mutans*. J Bacteriol, 184, 22, 6333-6342

Lis, M. & Kuramitsu, H.K. (2003). The stress-responsive *dgk* gene from *Streptococcus mutans* encodes a putative undecaprenol kinase activity. *Infect Immun*, 71, 4, 1938-1943

Loesche, W.J. (1986). Role of *Streptococcus mutans* in human dental decay. *Microbiol Rev*, 50, 4, 353-380

Moolenaar, W.H., Kruijer, W., Tilly, B.C., Verlaan, I., Bierman, A.J., & de Laat, S.W. (1986). Growth factor-like action of phosphatidic acid. *Nature*, 323, 6084, 171-173

Murayama, T. & Ui, M. (1987). Phosphatidic acid may stimulate membrane receptors mediating adenylate cyclase inhibition and phospholipid breakdown in 3T3 fibroblasts. *J Biol Chem*, 262, 12, 5522-5529

Nishizuka, Y. (1984). The role of protein kinase C in cell surface signal transduction and tumour promotion. *Nature*, 308, 5961, 693-698

Preiss, J., Loomis, C.R., Bishop, W.R., Stein, R., Niedel, J.E., & Bell, R.M. (1986). Quantitative measurement of *sn*-1,2-diacylglycerols present in platelets, hepatocytes, and *ras*- and *sis*-transformed normal rat kidney cells. *J Biol Chem*, 261, 19, 8597-8600

Qi, F., Merritt, J., Lux, R., & Shi, W. (2004). Inactivation of the *ciaH* Gene in *Streptococcus mutans* diminishes mutacin production and competence development, alters sucrose-dependent biofilm formation, and reduces stress tolerance. *Infect Immun*, 72, 8, 4895-4899

Quivey, R.G., Kuhnert, W.L., & Hahn, K. (2001). Genetics of acid adaptation in oral streptococci. *Crit Rev Oral Biol Med* 12, 4, 301-314

Raetz, C.R. & Newman, K.F. (1979). Diglyceride kinase mutants of *Escherichia coli*: inner membrane association of 1,2-diglyceride and its relation to synthesis of membrane-derived oligosaccharides. J Bacteriol, 137, 2, 860-868

Rampersaud, A., Harlocker, S.L., & Inouye, M. (1994). The OmpR protein of *Escherichia coli* binds to sites in the *ompF* promoter region in a hierarchical manner determined by its degree of phosphorylation. *J Biol Chem* 269, 17, 12559-12566

Rotering, H. & Raetz, C.R. (1983). Appearance of monoglyceride and triglyceride in the cell envelope of *Escherichia coli* mutants defective in diglyceride kinase. *J Biol Chem*, 258, 13, 8068-8073

Shibata, Y., van der Ploeg, J.R., Kozuki, T., Shirai, Y., Saito, N., Kawada-Matsuo, M., Takeshita, T., & Yamashita, Y. (2009). Kinase activity of the *dgk* gene product is involved in the virulence of *Streptococcus mutans*. *Microbiology*, 155, 557-565

Shibata, Y., Kawada-Matsuo, M., Shirai, Y., Saito, N., Li, D., & Yamashita, Y. (2011). *Streptococcus mutans* diacylglycerol kinase homologue: a potential target for anti-caries chemotherapy. *J Med Microbiol*, 60, 625-630

Tanzer, J.M., Livingston, J., & Thompson, A.M. (2001). The microbiology of primary dental caries in humans. *J Dent Educ* 65, 10, 1028-1037

Topham, M.K. & Prescott, S.M. (1999). Mammalian diacylglycerol kinases, a family of lipid kinases with signaling functions. *J Biol Chem*, 274, 17, 11447-11450

Walsh, J.P., Loomis, C.R., & Bell, R.M. (1986). Regulation of diacylglycerol kinase biosynthesis in *Escherichia coli*. A trans-acting *dgkR* mutation increases transcription of the structural gene. *J Biol Chem*, 261, 24, 11021-11027

Wen, Z.T. & Burne, R.A. (2004). LuxS-mediated signaling in *Streptococcus mutans* is involved in regulation of acid and oxidative stress tolerance and biofilm formation. *J Bacteriol*, 186, 9, 2682-2691

Wen, Z.T., Suntharaligham, P., Cvitkovitch, D.G., & Burne, R.A. (2005). Trigger factor in *Streptococcus mutans* is involved in stress tolerance, competence development, and biofilm formation. *Infect Immun*, 73, 1, 219-225

Wen, Z. T., Baker, H. V. & Burne, R. A. (2006). Influence of BrpA on critical virulence attributes of *Streptococcus mutans*. *J Bacteriol*, 188, 8, 2983-2992

Yamashita, Y., Takehara, T., & Kuramitsu, H.K. (1993). Molecular characterization of a *Streptococcus mutans* mutant altered in environmental stress responses. *J Bacteriol*, 175, 6220-6228

Yamashita, Y., Tsukioka, Y., Nakano, Y., Tomihisa, K., Oho, T., & Koga., T. (1998). Biological function of UDP-glucose synthesis in *Streptococcus mutans*. *Microbiology*, 144, 1235-1245

4

Human Milk: An Ecologically Functional Food

Adenilda Cristina Honorio-França and Eduardo Luzia França
Institute of Health and Biological Science at the Federal University of Mato Grosso, Barra do Garças, Mato Grosso, Brazil

1. Introduction

Breast milk is a valuable renewable resource that is often overlooked. Breastfeeding provides optimum nutrition for infant growth and development. It is considered the first line of defense for newborns because it is rich in soluble and cellular components that protect against gastrointestinal and respiratory infections (Hanson, 2001). Breastfeeding also has an economic value because it is one of the most efficient, cost-effective strategies to promote maternal and child health. In addition to being the safest and most natural way to feed a neonate, breastfeeding has been proven to protect neonates from a wide range of infectious and noninfectious diseases.

Extensive research using sophisticated epidemiologic methods and modern laboratory analyses has documented the diverse and compelling advantages conveyed by breastfeeding and the use of human milk for infant feeding. These advantages include health, nutritional, immunological, developmental, psychological, social, economic and environmental benefits (Honorio-França et al.,1997, Honorio-França et al., 2001, Hanson 2007, França et al., 2010, França et al., 2011a, França et al., 2011b). Several countries have published policy statements about breastfeeding and the use of human milk. Significant advances have subsequently occurred in both science and clinical medicine.

However, the ecological role of breastfeeding has not been extensively studied. Human milk is a unique food because it is produced and delivered to the consumer without pollution or wasteful packaging. It is a renewable resource and is beneficial in terms of nature conservation. Formula feeding, which is the substitute for breastfeeding, adversely affects the environment by depleting nonrenewable natural resources and causing damage at every stage of production, distribution and use (Hibbeln, 2002).

On the other hand, the environmental damage has been associated with the routine industrial and agricultural practices that are necessary for formula production. These industrial practices promote the destruction of areas covered by natural vegetation; waste electrical energy during the manufacturing process; accumulate a variety of packaging materials, such as aluminum and plastic; consume fuels for transportation and, finally, contribute to the expenditure of energy and water for formula preparation in millions of households (Krausmann, 2004).

Another important consideration is the presence of environmental contaminants in human milk. Some studies have reported high levels of environmental contamination in breast milk. These studies have focused on toxicants in human milk and their potential adverse effects on the breastfeeding child. However, the presence of environmental contaminants in human milk does not negate the advantages of breastfeeding (Lederman, 1996). Reduced breastfeeding would lead to increased infant mortality, increased demands on the health care system and an increased need for hard currency to buy more substitute infant foods, all of which would damage the environment. Rather than promoting formula feeding, public health professionals should mobilize the existing concerns about infant exposure to environmental contaminants through breast milk to help increase support for reducing the toxic environmental exposures of humans of all ages.

This chapter cites substantial new research on the importance of breastfeeding and describes the relationship between milk quality and various environmental, immunological, biochemical, pathophysiological and cultural factors. The chapter also discusses the environmental impacts of formula feeding and the environmental contaminants that affect milk quality and breastfeeding.

2. Cultural and behavioral aspects of breastfeeding

Breastfeeding has always been the gold standard for infant feeding. Throughout recorded human history, various populations recognized that breastfeeding was associated with reduced infant mortality, and evidence indicates that some populations did not survive due to artificially feeding their young (Eglash & Montgomert, 2008).

During the industrial revolution of the 20th century, many women left their children during the day to work in cities, and many children were fed not only with cow's milk but also a new product: infant formula. With pasteurization and refrigeration, the very high mortality rate of artificially fed infants declined, and artificial feeding became more popular.

Mothers assumed that formula feeding was a desirable modern option. Advertisements promoted the bottle as convenient and undermined the confidence of many mothers in their ability to feed their children. Although breastfeeding is considered instinctive, natural, and organic, some societal myths and taboos exist that should be abolished because they discourage mothers from breastfeeding their children. These myths include that the breasts sag or become deformed, stretch marks appear, the milk does not satisfy the infant, premature or low birth weight infants cannot be breastfed, the milk dries up, and breastfeeding mothers cannot exercise. The lack of knowledge among mothers led to them choosing not to breastfeed their children (Fujimori et al., 2008).

Although breastfeeding is a natural act, it is also a learned behavior. Research has shown that both mothers and health professionals need constant encouragement and support to maintain proper breastfeeding practices (Kent, 2007). Weaning is a social process, and as such, it should not be seen as an isolated, unicausal, timely act, except in very rare cases. Studies have shown that mothers and health care workers often cite the actual process of weaning as the "final cause". Weaning calls attention to the importance of "associated causes" (e.g., a "lactation crisis") that could lead to an interruption in breastfeeding for hours or days (Rea, 2004). Additional research should focus on the basic factors that trigger the weaning process and address the associated factors because without these factors, the terminal act does not occur (i.e., stopping the feeding of breast milk), and the weaning process is not entirely elucidated.

With increased scientific evidence of the risks of formula feeding and the incomparable benefits of breastfeeding, health organizations worldwide have published policy statements that affirm the importance of breastfeeding and the risks of artificial feeding for all populations. There is a growing movement among women and health professionals in many countries to re-establish infant feeding at the breast as the norm.

Breastfeeding babies and mothers are healthier, and breastfeeding has been shown to decrease health care costs for families, employers, and society. The education of health care professionals who work with mothers and infants regarding the dynamics of breastfeeding is necessary to allow them to effectively assist mothers. The recommendations suggest that every breastfeeding mother should have a health care professional observe at least one feeding while she is in the hospital or birthing center to ensure the proper transfer of milk to the baby and to decrease the risk of premature weaning (Bishara et al., 2008).

3. Maternal pathophysiological characteristics and breastfeeding: Influence during lactation

Mothers are the only source of nutrients for the fetus during development. Maternal nutritional deficiencies can engender a lack of critical nutrients during pregnancy with adverse consequences for both mother and infant.

The nutritional status of mothers should be evaluated during breastfeeding because it can affect the composition of the breast milk. It is particularly important for breastfeeding mothers to eat a well-balanced diet (Azizi & Smyth, 2009).

Deficiencies of iron and other minerals have been shown to have immunological and biochemical effects. The complexity of human milk makes it the ideal food source for babies for at least the first 6 months of their life. This early nutrition is an important environmental input that can exert lifelong effects on the metabolism and development of the child. The amount and composition of human milk is probably dependent of the diet of the mother (Shehadeh et al., 2006). Milk composition changes during lactogenesis and these changes can be used as biochemical markers of the onset of milk secretion. Importantly, normal infant development is sustained by the balanced gain of fat and calories provided by breastfeeding (França et al., 2010). Human colostrum is rich in biologically active molecules, which are essential for antioxidant functions. The soluble components of these molecules act in a child's gut without provoking an inflammatory response.

The relationship between the components of human milk and the maternal pathophysiological characteristics is a matter of debate. Some studies have suggested that maternal characteristics cause variations in the components of colostrum and milk, whereas others have reported that these changes in milk composition are caused by nutritional and disease maternal status (AAP, 2005, Zhanga et al., 2010).

Studies have attempted to elucidate the effects of lactation on maternal glucose metabolism. In this chapter, we address the effects of maternal hyperglycemia on the biochemical and immunological composition of colostrum (Morceli et al., 2011).

Colostrum has a low protein concentration in diabetic women, whereas in normoglycemic women, the protein levels are within the reference limits. In fact, protein and glucose levels in the colostrum of diabetic women are likely maintained at normal levels and at a constant ratio with capillary concentrations, which suggests that adequate glycemic control can correct any abnormalities in milk composition (Morceli et al., 2011, França et al., 2011b). The improper control of glycemia may have undesirable consequences, such as compromised breastfeeding due to a delayed lactogenesis transition from phase I to phase II (Oliveira et al., 2008).

The production of milk components changes in diabetic mothers due to alterations in glucose metabolism. Therefore, adequate maternal glycemic control is crucial in diabetic mothers to ensure both that the nutritional needs of their newborns are met and that the immunologic components are properly provided. Despite the potential abnormalities in the breast milk of women with diabetes or nutritional deficiencies, these women should be strongly encouraged to breastfeed their children (Morceli et al., 2011, França et al., 2011b). Not only is breast milk an excellent food source for newborns, it also decreases the high rates of maternal and infant complications. In addition, the growth rate of breastfed infants is related to the total amount of milk they consume rather than the concentration of fat, proteins or carbohydrates in the milk.

The importance of both maternal health and the role of breastfeeding in child development necessitate the development of new public health policies to promote nutrition education for mothers at risk of disease and to ensure high-quality milk for their children.

4. Influence of intrinsic factors on breastfeeding

The volume of breast milk produced does not vary significantly between population groups, with mothers in developing countries producing as much milk as mothers in developed countries.

Breastfeeding is economically valuable because it is one of the most efficient, cost-effective strategies for providing health benefits to both mothers and newborns. The composition of breast milk is dynamic, and it is influenced by intrinsic factors. The chemical composition of breast milk changes over time according to the nutritional needs of the infant; i.e., it adjusts to pregnancy stage (prepartum and postpartum) and time of day (night and day). The breast milk produced for premature infants has more protein and additional factors to protect the infant until the 30th postpartum day (Delneri et al., 1997). Breast milk can be classified as early lactation (260 days of gestation), colostrum (1 to 7 days postpartum), transitional milk (7 to 15 days postpartum) and mature milk (after 15 days postpartum - Kent, 2007).

Colostrum is a yellow, viscous fluid rich in immunoglobulins, proteins, fat-soluble vitamins, minerals and leukocytes. Transitional milk is rich in proteins, enzymes, vitamins and minerals but has a lower leukocyte concentration. Mature milk consists mainly of water, which is sufficient for the needs of the child, in addition to proteins, carbohydrates, lipids, minerals and vitamins.

The constitutional differences of breast milk in the postpartum period suggest that this secretion shows chronobiological variation. Breastfeeding is important to ensure an adequate passive transfer of immunity and intake of nutritional components. Human colostrum and breast milk change as a function of both time of day and milk maturation, and this variation is important to mothers and to those responsible for the collection and distribution of milk in a human milk bank (França et al., 2010).

Intrinsic factors are responsible for the daily fluctuations in milk composition that increase progressively up to 3 months postpartum, and one theory posits that one of the roles of such rhythms in milk composition is to stabilize the circadian system of the newborn at a time when other means of stabilization (i.e., those functioning in adults), such as the rest-activity cycle, have not yet developed fully.

The temporal influence on circadian rhythms in the cellular, immunological and biochemical components of human milk suggests that human milk presents chronobiological variation that is influenced by changes during the postpartum period. The adjustment to pregnancy stage (prepartum and postpartum) and time of day (night and day) in the composition of human milk suggests an auxiliary physiological mechanism for the defense of the infant against infections because there is a predominance of immunological components in milk in the daytime when infants are most vulnerable due to contact with infectious agents. This variation might represent both an additional breastfeeding mechanism to improve newborn adaptation to environmental changes and an endogenous force promoting the establishment of biological rhythmicity in humans (França et al., 2010).

5. Immunological and environmental factors in human milk

In considering the immunological significance of the infant immune system, the maternal immune system, and the interaction between the two, various concepts and models must be considered, including innate and adaptive immunity, mucosal immunity, inflammatory and anti-inflammatory responses, active versus passive immunity, dose-response relationships, and the dynamic nature of acute immune responses.

The mucosal epithelia of the gastrointestinal, upper and lower respiratory, and reproductive tracts cover a surface area over 200 times greater than that of the skin. These surfaces are especially vulnerable to infection due to their thin, permeable barriers. The mucosal surface has many physiologic functions. The most important function of the collective mucosal surfaces is immunological: protection against microorganisms, foreign proteins, and chemicals as well as immune tolerance to many harmless environmental and dietary antigens.

Human milk has mucosal immunologic components that can change or interact with the environment and provide benefits for the infant (Brandzaeg, 2010). The most important contribution to the dynamic nature of breast milk is the mucosal associated lymphoid tissue (MALT) system. When an infant and mother are exposed to a potential pathogen within their environment, the mature maternal immune system can react more quickly and effectively than that of the infant. It is the dynamic nature of breast milk, with all its bioactive factors, and the interaction between the infant and maternal immune systems through breast milk that makes human breast milk a truly unique, incomparable, and ideal source of nutrition for infants.

The composition of milk is modified during feeding and according to any diseases experienced by the mother. Newborns lack the ability to launch effective immune responses against microorganisms, and for several months after birth, the primary defense against infections is the passive immunity provided by maternal antibodies. Colostrum and human milk have been thoroughly studied in recent years, ever since their protective role against infections was confirmed (Honorio-França et al., 1997, Honorio-França et al., 2001, França-Botelho et al., 2006, Lawrence & Pane, 2007, França et al., 2011a, França et al., 2011b). Human milk was first used clinically as a vehicle for the transfer of passive immunity, but its immune components are now known to be highly immunoreactive, exhibiting time-dependent alterations.

Human milk contains several immunoreactive proteins (e.g., IgA, IgG, IgM, C4, C3 and others) but is particularly rich in secretory IgA (SIgA). SIgA protects against a number of microorganisms by acting as an opsonin, blocking bacterial adherence to epithelial cells, neutralizing toxins and preventing viral infections (Honorio-França et al., 1997, Hanson, 2007). These functions are complemented by the antibodies IgM and IgG and particularly by complement proteins C3 and C4. SIgA in breast milk serves as an optimal antigen-targeted passive immunization of the gut of the breastfed infant. Breastfeeding is therefore the best defense against mucosal infection in developing countries. The protection offered by breast milk depends not only on levels of immunoglobulin or other immunoreactive proteins but also on the amount, timing and type of milk consumed by the infant.

In addition to antibodies, soluble bioactive components and anti-infectious factors, human colostrum also contains large amounts of viable leukocytes (10^9 cells / mL in the first days of lactation), especially macrophages and neutrophils. These cells produce free radicals and have phagocytic and bactericidal activity (Honorio-França et al., 1997, Honorio-França et al., 2001, França et al., 2011a, França et al., 2011b). In bacterial infections, phagocytes are known to be the main cell lineage in the host's defense.

Colostral phagocytes exhibit phagocytic activity and may be activated by stimulatory signals generated by milk antibodies, especially during the interaction between SIgA and its Fc receptor. These cells produce oxygen free radicals and present bactericidal activity after opsonization with SIgA with a rate equivalent to that of phagocytes from peripheral blood (Honorio-França et al., 2001; França et al., 2011b).

The significance of the biological activity of the soluble and cellular components present in human milk is of great importance because human milk contains the highest amount of these immunoreactive proteins and may represent a complete micro-environment in which both soluble and cellular components act together.

The rich nutrition provided by breast milk in early nutrition has lifetime beneficial effects on metabolism, growth, immunity, neurodevelopment and major disease processes. Breast milk also provides additional protection against intestinal and respiratory infections in infants. Mothers should therefore be encouraged to breastfeed their babies because this is an important strategy for prevention newborn infections.

Recently, attention has been focused on the environmental damage associated with routine industrial and agricultural practices. A developing ecology movement has stimulated scientific concern about these problems and focused attention on the possible association between breastfeeding and the environment.

Contaminants can influence the quality of human milk. Studies have reported the presence of persistent organic pollutants in particular dioxins and furans (Krausmann, 2004). These organic pollutants are created and released into the environment through industrial chemical processes, and they have been described as highly toxic substances capable of contaminating water, air, and soil. When these pollutants enter the food chain, they become a serious public health problem (Lederman et al., 1996).

Due to their high liposolubility, these organic pollutants accumulate in the fatty tissues of living organisms and deposit on the surface of plants, fruits, and vegetables, which are then ingested by humans and animals. The pollutants accumulate in fatty tissues and subsequently enter the breast milk. These pollutants can be transferred to the child through breastfeeding and have serious implications, including potentially carcinogenic effects.

Concerns about breast milk contamination have contributed to declining rates of breastfeeding. This reduction in breastfeeding has engendered an increase in the use of human milk substitutes. The increased use of Formula-fed has been associated with deforestation, erosion, pollution, climate change and waste materials. Promoting premature weaning contributes to the destruction of natural resources.

The process of industrial modernization is characterized by fundamental changes in the interaction between socioeconomic systems and their natural environments (Ferrari, 2007).

The use of products such as industrially produced baby food has increased the environmental impact.
Breastfed children and formula-fed children experience different rates of morbidity and mortality. There is evidence of short- and long-term benefits of breastfeeding. Artificially fed infants have significantly higher rates of acute otitis media, nonspecific gastroenteritis, severe lower respiratory tract infections, atopic dermatitis, asthma, sudden infant death syndrome (SIDS), and necrotizing enterocolitis than breastfed infants.
Breastfeeding also contributes to a decrease in the use of industrialized milk, bottles and other products that damage the environment. Breastfeeding is an ecological act because it contributes to the environment by promoting an environment for current and future generations that enhances their quality of life and lowers their risk of malnutrition and infection.
The importance of breastfeeding to child development necessitates the development of new public health policies to support decreased levels of environmental contaminants for mothers living in high-risk areas and to ensure high-quality breast milk for their children.

6. Conclusion

Although both rates of breastfeeding and the composition and quality of human milk are influenced by cultural, behavioral, pathophysiological, intrinsic, immunological and environmental aspects, women should be strongly encouraged to breastfeed their children. In addition to being an excellent food source for newborns, breast milk decreases the high rates of maternal and infant complications. In addition, the growth rate of breastfed infants is related to the total amount of milk they consume rather than the concentration of fat, proteins or carbohydrates in the milk because breast milk is the most ecologically sound source of nutrition.
Breastfeeding is an ecological act because it supports an environment for current and future generations that enhances their quality of life and lowers their risk of malnutrition and infection.

7. References

American Academy of Pediatrics (AAP). Breastfeeding and the use of human milk. Pediatrics. 115:492-506, 2005.

Azizi F, Smyth P. Breastfeeding and maternal and infant iodine nutrition. Clin Endocrinol. 70: 803–809, 2009.

Bishara R, Dunn MS, Merko SE, Darling P. Nutrient composition of hind milk produced by mothers of very low birth weight infants born at less than 28 weeks'gestation. J Hum Lact 24:159-167, 2008.

Brandtzaeg P. The mucosal immune system and its integration with the mammary glands. J Pediatr. 156:S8-15, 2010.

Delneri MT, Carbonare SB, Silva MLM, Palmeira P, Carneiro-Sampaio MMS. Inhibition of Enteropathogenic *Escherichia coli* adhesion to Hep-2 cells by colostrum and milk from mothers delivering. Eur J Pediatr. 156: 493-498, 1997.

Eglash A, Montgomery A, Wood J. Breastfeeding. Dis Mon. 54:343-411, 2008.

Ferrari C. Functional foods and physical activities in health promotion of aging people. Maturitas. 327 - 339, 2007.

França E L, Nicomedes TR, Calderon IMP, Honorio-França AC. Time-dependent alterations of soluble and cellular components in human milk. Biol Rhythm Res. 41:333-347, 2010.

França EL, Bitencourt RV, Fujimori M, Morais TC, Calderon IMP, Honorio-França AC. Human colostral phagocytes eliminate enterotoxigenic *Escherichia coli* opsonized by colostrum supernatant. J. Microbiol Immunol Infec.44:1-7, 2011

França EL, Morceli G, Fagundes DLG, Rudge MVC, Calderon IMP, Honorio-França AC. Secretory IgA- Fcα Receptor interaction modulating phagocytosis and microbicidal activity by phagocytes in human colostrum of diabetics. Acta Pathol Microbiol Immunol Scand (119: 710-719), 2011.

França-Botelho AC, Honorio-França AC, França EL, Gomes MA, Costa-Cruz JM Phagocytosis of *Giardia lamblia* trophozoites by human colostral leukocytes. Acta Paediatr. 95:438-443,2006.

Fujimori M, Morais TC, França EL, de Toledo OR, Honório-França AC. The attitudes of primary school children to breastfeeding and the effect of health education lectures. J Pediatr (Rio J). 84:224-231, 2008.

Hanson LA Feeding and infant development breast-feeding and immune function. Proc Nutrit Soc. 66: 384-396, 2007.

Hibbeln JR. Seafood consumption, the DHA content of mothers' milk and prevalence rates of postpartum depression: a cross-national, ecological analysis. J Affect Disord. 69:15-29, 2002.

Honorio-França AC, Carvalho MP, Isaac L, Trabulsi LR, Carneiro-Sampaio MMS. Colostral mononuclear phagocytes are able to kill enteropathogenic *Escherichia coli* opsonized with colostral IgA. Scand J Immunol. 46:59-66,1997.

Honorio-França AC, Launay P, Carneiro-Sampaio MMS, Monteiro RC. Colostral neutrophils express Fc alpha receptors (CD89) lacking gamma chain association and mediate noninflammatory properties of secretory IgA. J Leukoc Biol. 69:289-296, 2001.

Kent JC. How Breastfeeding Works. J Midwif Women Health. 52: 564-570, 2007.

Krausmann F. Milk, manure, and muscle power. Livestock and the transformation of preindustrial agriculture in Central Europe. Human Ecology. 32:735-772, 2004.

Lawrence RM, Pane CA. Human Breast Milk: Current Concepts of Immunology and Infectious Diseases. Curr Probl Pediatr Adolesc Health Care.37:7-36, 2007.

Lederman SA. Environmental contaminants in breast milk from the central Asian Republics. Reproduct Toxicol. 10: 93:104, 1996.

Morceli G, França EL, Magalhães VB, Damasceno DC, Calderon IMP, Honorio França AC. Diabetes induced immunological and biochemical changes in human colostrum. Acta Paediatr. 100: 550-556, 2011.

Oliveira AMM, Cunha CC, Penha-Silva N, Abdllah VOS, Jorge PT. Interference of the Blood Glucose Control in the Transition Between Phases I and II of Lactogenesis in Patients with Type 1 Diabetes Mellitus.Arq Bras Endrocrinol Metab. 52:473-81, 2008.

Rea MF. Benefits of breastfeeding and women's health. J Pediatr (Rio J). 80:142-146, 2004.

Shehadeh N, Aslih N, Shihab S, Werman M J, Sheinman R, Shamir R. Human milk beyond one year post-partum: lower content of protein, calcium, and saturated very long-chain fatty acids. J Pediatrics.148:122-124, 2006.

Zhanga Y, Lib N, Yanga J, Zhanga T, Yang Z. Effects of maternal food restriction on physical growth and neurobehavior in newborn Wistar rats. Brain Research Bulletin. 83:1–8, 2010.

Part 3

Environmental Monitoring

Factors Controlling the Incorporation of Trace Metals to Coastal Marine Sediments: Cases of Study in the Galician Rías Baixas (NW Spain)

Belén Rubio[1], Paula Álvarez-Iglesias[1], Ana M. Bernabeu[1], Iván León[2], Kais J. Mohamed [1], Daniel Rey[1] and Federico Vilas[1]
[1]Universidad de Vigo, Vigo, Pontevedra
[2]Universidad del Atlántico, Barranquilla
[1]Spain
[2]Colombia

1. Introduction

Transitional coastal environments such as the Galician Rías in the Atlantic coast of NW Spain are densely populated areas. Their environmental problems are highlighted by the conflicting interests of different economic sectors: extensive mariculture activities are located in its waters and intertidal zone; shipbuilding, carbuilding, canning and other industries compete with tourism on their shores; and dairy farming is the main agricultural activity in its surrounding hills and hinterland (Vilas et al., 2008). As a result, the management of the coastal zone is highly complex and it is difficult to balance quality of coastal waters with economic activities. For instance, in the Ría de Vigo, the southernmost of the Rías Baixas, wastewater treatment plants were not installed until the 1990s, and in spite of regional environmental legislation (Lei 8/2001), their capacity was still insufficient in 2005 when the European Court of Justice found Spain guilty of failure to fulfill its obligations under the Article 5 of the Council Directive 79/923/EEC on the quality required for shellfish waters (Case C-26/04 ECJ). This case was closed following Spain's submission of a pollution-reducing programme specifically pertinent to shellfish waters; success of this plan will depend critically on the behaviour of the sediments on the ría bottom.

Galician Rías experience seasonal upwelling, which increases marine productivity. This promotes the deposition of high organic matter contents in the bottom sediments and contributes to the observed intense sedimentation rates of 1-6 mm yr^{-1} (Álvarez-Iglesias et al., 2007; Rubio et al., 2001). Current levels of trace metals (Prego & Cobelo, 2003) in sediments of these Rías have caused a significant concern by local and European authorities in the last ten years, especially in relation to the application of the Water Framework Directive (WFD), aimed to ensure that all waters reach "good status" by 2015. Some of these studies (Álvarez-Iglesias et al., 2003; Belzunce-Segarra et al., 2008; Rubio et al., 2000a) showed that the highest concentration of trace metals occurs in the muddiest surficial

sediments of the rías, and that their fate and bioavalability depends on the intensity and speed of bacterial-mediated redoxomorphic post-sedimentary processes (Álvarez-Iglesias & Rubio, 2008, 2009; Rubio et al., 2010). This chapter will review the main factors that control the incorporation of metals to the sediments in these environments, with a focus on the forcing factors and their temporal evolution in the recent sedimentary record.
This study will also show the critical importance of distinguishing and quantifying the various metal forms by using sequential extractions and by the determination of magnetic properties in order to reach a full understanding of the potential and present environmental impacts of contaminated sediments. Special emphasis will be put in the role of mussel rafts on the diagenetic inmobilization of heavy metals. Finally, the solubility of these metals by aerobic oxidation will be analyzed in some laboratory experiments in order to improve coastal risks prevention and management.

2. Sediments as trace metal sinks and sources

Water analyses proposed in the WFD (Directive 2000/60/CE) are the most obvious way to quantify the degree of metal contamination in an area. However, these analyses are not easy because concentrations of metals in solution are very low, contamination can occur during collection and analysis, and sampling needs to be repeated in specific time intervals (weeks, months, and seasons). Moreover, most metals transported in aquatic ecosystems quickly set on the solid material, due to their low solubility (Forstner & Wittman, 1981). Binding of metals in suspension will eventually lead to their incorporation into the sediment (Fig. 1).

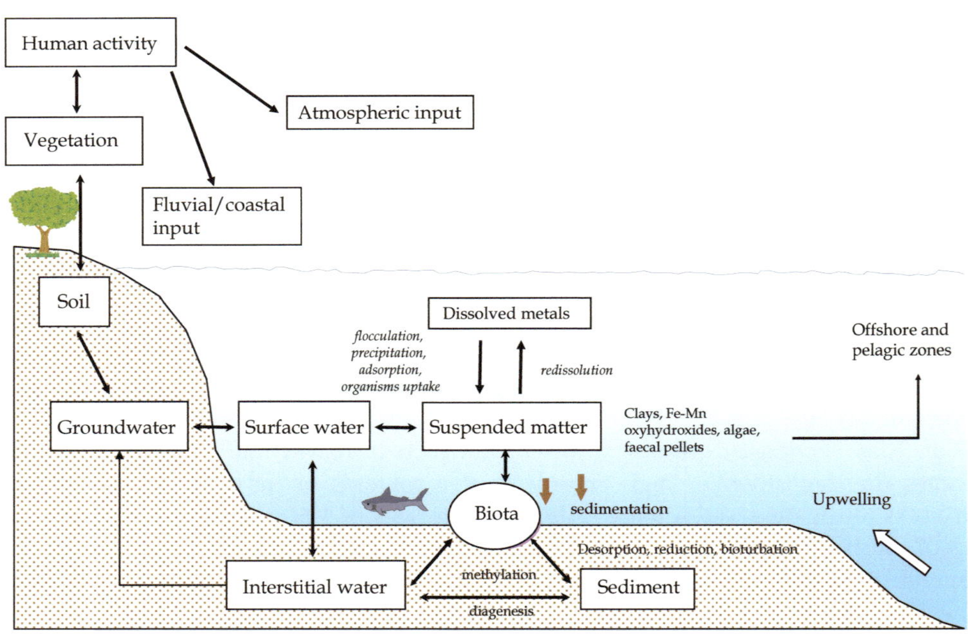

Fig. 1. Schematic representation of metals reservoirs and their interactions in coastal systems.

Therefore, sediments are a sink of metals with concentrations several orders of magnitude higher than those in the adjacent waters, both interstitial and overlying (Tessier & Campbell, 1988). Nevertheless, sediments are dynamic reservoirs subject to rapidly changing conditions. When the environmental variables change, remobilization of metals can occur. Although there are different mechanisms of metal binding to sediments, adsorbed metals appear to be more readily available, and therefore can be recycled. In these cases, the sediment acts as a source of metals to other biotic and abiotic compartments (Fig. 1).
In addition, sediments can be considered as archives of environmental information due to their "memory capacity", so that the sedimentary record allows us to reconstruct the recent historical record of coastal pollution (Álvarez-Iglesias et al., 2007; Rubio et al., 2001, 2010; Valette-Silver, 1993; among others).

3. Incorporation of metals to the sediment

Human activities have drastically altered the biogeochemical cycles and equilibria of some trace metals. These metals cannot be degraded or destroyed and become stable and persistent contaminants that tend to accumulate in sediments. Metals can be transferred from sediments to benthic organisms and then become a potential risk to human consumers by incorporation through the food web (Soto-Jiménez et al., 2011).
The main anthropogenic metals sources are industrial point sources, including present and former mining activities, foundries and smelters, shipbuilding, chemical industries, metallic industries, and diffuse sources such as combustion by-products. Dispersion of metals in the particulate phase is usually small, but relatively volatile metals and those that become attached to air-borne particles can be widely dispersed over very large scales. Trace metals carried in dissolved or particulate forms (e.g., river run-off) enter the normal coastal biogeochemical cycle and are largely retained within near-shore and shelf regions (Fig. 1).

3.1 Processes affecting the cycles of metals in coastal zones

Trace elements may suffer varying degrees of internal recycling before they are buried in the sediment and/or carried into the ocean (Fig.1). Such recycling may involve processes such as flocculation, precipitation, release from living or dead particulate phases, and subsequent regeneration when these particles undergo redissolution. Recycling of metals in suspended solids takes place by coprecipitation, adsorption, desorption and flocculation (Fig. 1). The suspended matter and deposited sediments are linked through processes of sedimentation and erosion. Diagenetic processes release high concentrations of trace metals to interstitial waters, which can influence metal concentrations in the overlying waters through diffusion, consolidation and bioturbation. This element recycling can occur within the water column or within the sediment. If the metal residence time is short an element can be recycled several times.
In addition to physical processes, recycling in the sediment can also be biologically mediated (i.e. methylation) (Fig. 1). The concentration of suspended matter may also influence these processes, especially in estuaries and rías, where suspended matter concentration is much larger than in other systems of the hydrologic cycle, such as most lakes and oceans.

3.2 The interaction between trace metals and aquaculture

In the last decades, marine aquaculture has experienced an important development around the world. Galicia is the second largest producer of mussels in the world after China. They are cultivated in wood frames called mussel rafts. Most of these are concentrated in the Rías Baixas, with more than 3,000 rafts located in Arousa, Pontevedra and Vigo rías.

An important environmental impact of these activities is the high amount of particulate matter discharged by mussels from faeces. Although the concentration of heavy metals in these particles is relatively low, the amount of solids is so high that the total accumulation of metals in the sediments may become an important problem. This fact has been mentioned in previous works (Otero et al., 2005; Prego et al., 2006), but it has not been studied in depth. Table 1 compares the accumulation of some trace metals (especially Pb, Ni, and V) in sediments collected below mussel rafts and in adjacent areas in the Ría de Pontevedra. Despite intensive marine aquaculture these results indicate that the differences are not very high, and they seem to be more related to textural differences than to aquaculture activitities. However, there are very significant differences in the elements and ratios of the organic matter characterization (table 1). Sediments collected below mussel rafts areas showed higher contents of total organic carbon (TOC), total N (TN) and total S (TS) than those collected in adjacent areas (table 1). Significant differences were also observed in the mean values for the ratios C/N and S/C, showing that the increase in TOC in mussel rafts areas influences the redoxomorphic organic matter degradation. C/N ratios are, on average, higher than those reported for biodeposits by other authors (<10; Calvo de Anta, 1999; Otero et al., 2006). S/C ratios are below the global average in normal marine sediments (Raiswell & Berner, 1986), indicating a moderate stage of diagenetic evolution.

Trace elements ($\mu g\ g^{-1}$)	Mussel rafts areas	Adjacent areas
Sr	865 ± 504	1888 ± 438
Rb	210 ± 31	169 ± 41
Ba	403 ± 73	341 ± 58
Co	13 ± 2	14 ± 1
Cu	20 ± 7	29 ± 20
Zn	77 ± 20	89 ± 32
Ni	34 ± 12	24 ± 4
Pb	18 ± 15	3 ± 8
Cr	65 ± 12	66 ± 11
V	89 ± 29	65 ± 17
Other parameters (%)		
TOC	3.69 ± 1.76	2.42 ± 0.68
TN	0.24 ± 0.13	0.13 ± 0.77
TS	0.95 ± 0.62	0.37 ± 0.16
C/N	19.83 ± 10.18	25.19 ± 13.33
S/C	0.25 ± 0.13	0.16 ± 0.09

Table 1. Comparison of trace elements concentration obtained by X-ray Fluorescence (XRF) and other sediment parameters (TOC, TN, TS, C/N and S/C) for a group of sediment cores (101 samples) collected below mussel rafts and in adjacent areas (35 samples) in the Ría de Pontevedra.

4. The ría environment - Factors controlling trace metal contents in ría sediments

The Rías Baixas are a characteristic geomorphological coastal feature of the Northwest Iberian Margin consisting of four deep and narrow V-shaped Tertiary river valleys that have been flooded during the last sea-level transgression.

Most regional studies in rías have shown that although the hydrodynamic processes are similar to those identified in estuaries, the rías are clearly dominated by the waves while the estuarine circulation is restricted to the innermost areas (Piedracoba et al., 2005; Ruiz-Villarreal et al., 2002; Souto et al., 2003; Vilas et al., 2005). These environments are also characterized by a lesser continental freshwater input, and a higher primary productivity due to seasonal upwelling (Fraga, 1981) in comparison to estuarine environments.

In addition, the sediment characteristics and distribution of the Galician Rías Baixas (Vilas et al., 2005) also show significant differences from the facies models of wave- or tide-dominated estuaries (Vilas et al., 2010) as we will discuss in the following sections.

4.1 Factors and forcings controlling grain-size distributions

Wave conditions exert an important control on sediment distribution (Rey et al., 2005; Vilas et al., 2005, 2010). Organic-rich fine-grained sediments accumulate in low-energy areas along the deep central axis, and in protected areas of the inner ría sector with maximum mud percentages near 100% (Fig. 2).

Mud accumulation is also promoted by the agglutinating effect of organic matter. As a result, organic matter content is higher in muds, and increases towards the inner ría to values in excess of 10% (Vilas et al., 2005).

Sediment composition inside the rías is predominantly siliciclastic, as a result of the granitic and metamorphic rocks of their catchment areas. As an example, figure 2 shows the similarities between quartz distribution and mud contents. On the contrary, biogenic carbonates are predominant in the sand and gravel fractions. Production of these coarse calcareous bioclastic sediments is favoured by upwelling fertilization of the rías. $CaCO_3$ abundance is greatest at the margins of the ría and towards the outer areas (Fig. 2), where wave energy is stronger. In these areas $CaCO_3$ contents can reach values higher than 90%.

Many authors have recognized the sediment grain size as a factor directly related to the ability for retaining trace elements (Horowitz & Elrick, 1987). This relationship is clearly observed in the sediments of the rías by the surficial distribution of Pb (Fig. 2) and other trace elements. This is also shown by the strong positive correlations between mud percentage and trace elements concentrations (Fig. 3). This correlation is explained by a combination of physico-chemical factors, since materials with a higher capacity to retain contaminants have smaller particle sizes and therefore also have higher specific surfaces and cation exchange capacities.

In addition, the effect of grain size is enhanced by organic matter, which is a complexing agent for some pollutant metals and is concentrated in fine-grained particles (Wangersky, 1986). Note in figure 3 the typical association of Pb and Cu with organic matter and the strong relationship of Co with finer fractions. The diluent effect, expressed as a negative correlation, caused by coarser fractions and/or carbonates, is exemplified by the concentration of Zn vs the percentage of $CaCO_3$ or the concentration of Cu vs the percentage of sand.

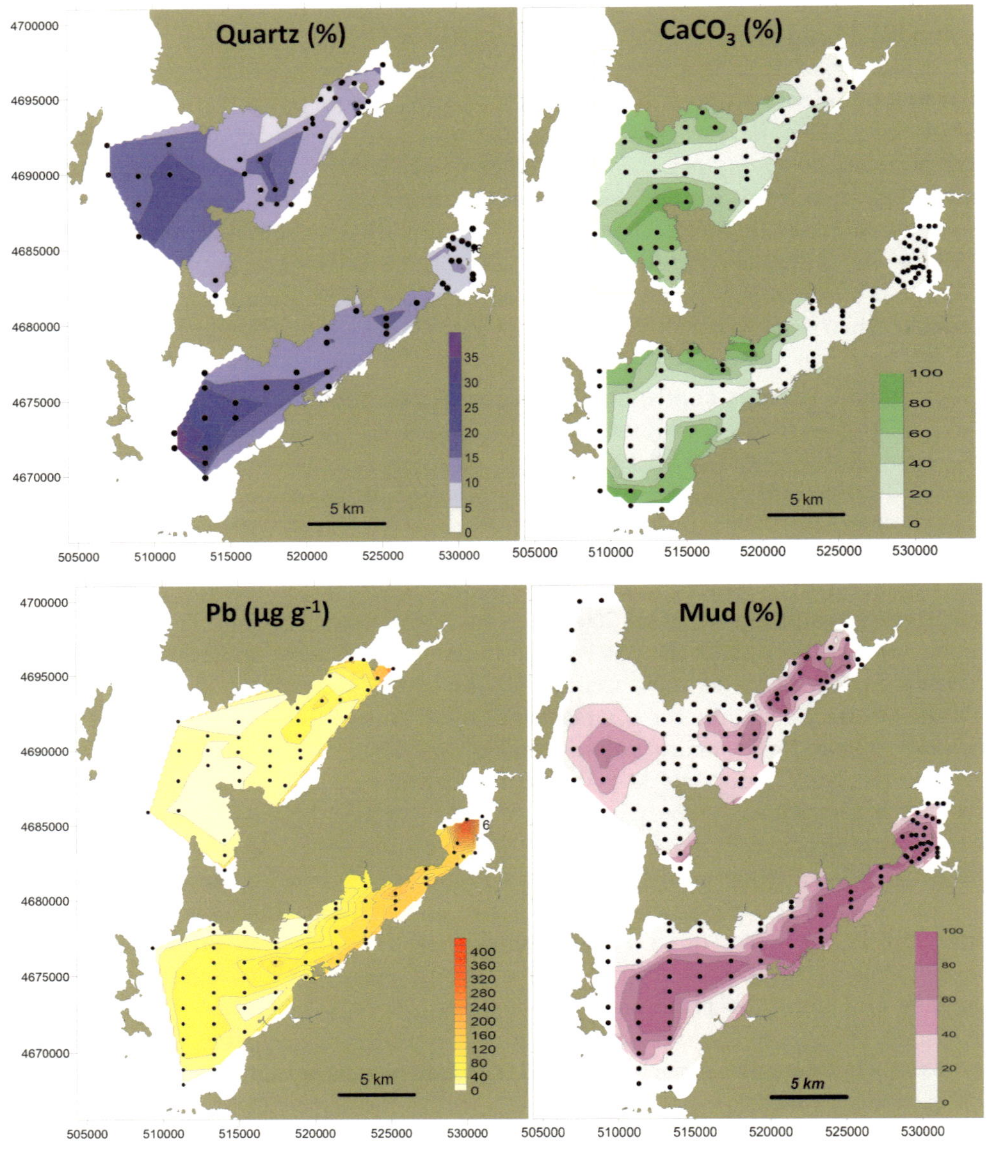

Fig. 2. Distribution maps of quartz, calcium carbonate, lead and mud concentrations in surface sediments of the Rías de Vigo and Pontevedra (NW Spain) measured on more than 100 samples (black dots).

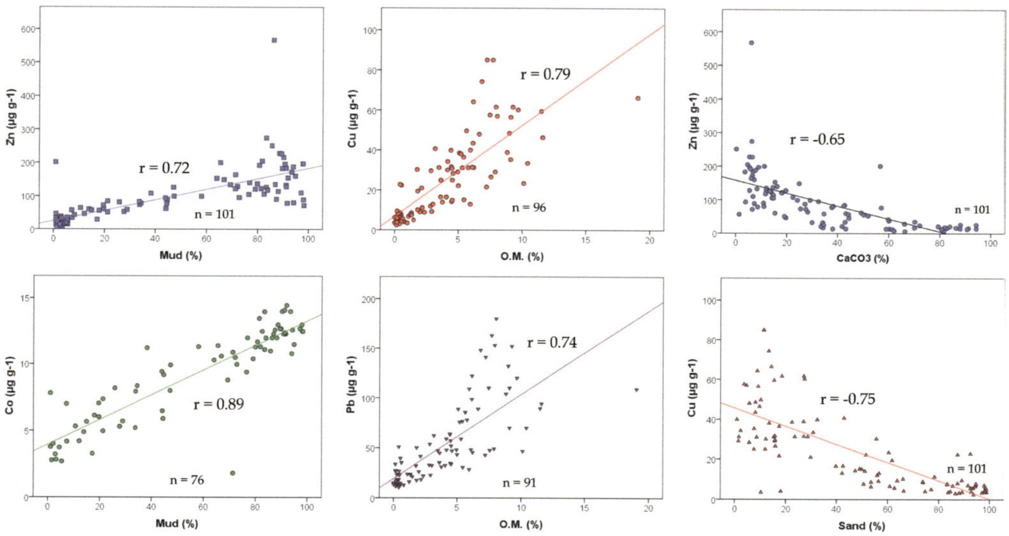

Fig. 3. Relationships between some trace elements and some properties of about one hundred of surficial sediment samples located in the Rías de Vigo and Pontevedra (data from Ría de Vigo from Rubio et al., 2000a). Sample location in figure 2 (black dots).

4.2 Grain-size effect: Proxies and normalization procedures

A very simple method used to detect whether a sediment is contaminated is to map the surface concentration of the target element and try to detect geochemical anomalies (Chester & Voutsinou, 1981) that highlight areas or regions with anomalous contents. For instance, too high values of Pb were detected in the inner part of the Ría de Vigo (San Simón Bay) (Fig. 2). In addition, the distribution patterns of conservative elements indicative of grain size should be compared to the distribution of trace metals in order to detect whether or not these metals are supplied by anthropogenic activities.

However, a first approach to determine the presence of contamination is to analyze the relationships between a normalizer element or grain-size proxy (Al, Ti, Rb, among others) and the potential contaminant element. If there is no linear relationship between them, this is usually due to contamination. For example for the relationship between Zn and mud (Fig. 3), those data points that are far from the correlation line are indicative of contamination.

5. Anthropogenic evidences on metal concentration in ría sediments

Several indexes (contamination factor, enrichment factor, geoaccumulation index, among others) have been developed to assess the degree of metal contamination in a given area. These indexes compare the metal content of the samples with natural values for each metal. The determination of these so-called background levels is a key factor in assessing the degree of contamination or the anthropogenic effect in a given area. Rubio et al. (2000a) showed how the choice of these values determines the geochemical interpretation of a given area, hence the importance of establishing background values adequately.

Also the comparison with Sediment Quality Guidelines (SQGs) that allows calculating the effects range low (ERL), effects range medium (ERM) and probable effect levels according to

Long et al. (1995) has been used by several authors (Mucha et al., 2003; Pekey et al., 2004). In the following sections we will review some examples for the sediments of the Rías Baixas.

5.1 Background levels

The background value or "background" of a given trace metal in sediments is the natural content of the metal without human intervention. This value will depend on the geochemistry of the source area sediment. Several possibilities have been set up to establish background values for trace metals (Forstner &Wittmann, 1981):

1. Mean values of metals in the crust (Taylor, 1964) or average shale values (Turekian & Wedepohl, 1961; Wedepohl, 1971, 1991).
2. Values determined by various methods, in the same study area, including:
   a. Selection of presumably clean stations (Barreiro et al., 1988; Subramanian & Mohanachandran, 1990).
   b. Statistical methods, among others, include: multiple regression techniques (Summers et al., 1996a), principal component analysis (Rubio et al., 2000), selection of the first percentile of the cumulative distributions of the concentration of metals (Barreiro et al., 1988), and determination of homogeneous populations based on the analysis of frequency distribution curves (Carral et al., 1995).
3. Analysis of sediment cores deep enough to contain the preindustrial record in the sediment (Angelidis & Aloupi, 1995), which is the best recommended technique for establishing background values for a particular area. For example, Rubio et al. (2000b) proposed background values for the Ría de Vigo from a core about 3 m long, with an approximate age of over 1000 years BP enough to reach preindustrial levels (Table 2).

Table 2 gives some examples of background concentrations obtained for several authors for typical trace metals found in the rías compared with global background values. In many cases background values at the global level can be inadequate for a particular area and it is necessary to obtain background values at local or regional level.

Metal	B (1)	C (2)	R(3)	R (4)	A (5)	T (6)
Al	--	--	6.48	6.48	9.82	8.0
Fe	2.69	2.95	3.51	3.51	3.53	4.72
Ti	--	--	0.34	0.34	0.36	0.46
Mn	225	273	244	244	216	850
Zn	100	133	105	105	110	95
Cu	25	22	29	20	21	45
Pb	25	73	51	25	51	20
Cr	43	34	34	55	65	90
Ni	30	32	30	30	33	68
Co	16	12	12	12		19

(1) Barreiro 1991. (2) Carral et al., 1995. (3) Rubio et al., (2000a). (4) Rubio et al. (2000b). (5) Álvarez-Iglesias et al., 2006. (6) Average shale values from Turekian & Wedepohl (1961).

Table 2. Regional background values obtained by different authors for ría sediments, and its comparison to average shale values from Turekian & Wedepohl (1961). Shadowed values are similar between the different authors.

5.2 Studies on sediment cores in the rías: The need of dating with ^{137}Cs and ^{210}Pb

During the last decade, radionuclide dating of sediment cores has been used to establish sources and input rates of pollutants such as trace metals (Lee & Cundy, 2001; Ligero et al., 2002). However they have been very rarely used in sediments for the Galician Rías (Álvarez-Iglesias et al., 2007; Rubio et al., 2001). Among the latest methods to determine these rates the depth distribution of ^{210}Pb and ^{137}Cs specific activities have proven to be valid. ^{137}Cs is a good tracer for erosion and sedimentation because there are no natural sources of this radioisotope that is produced during nuclear fission. Its presence in the environment, therefore, is due to nuclear testing or release from nuclear reactors. The distribution of this radioisotope in a sediment core would reflect variations in their inputs to the environment. Average sedimentation rates are obtained by identifying their maximum inputs in the activity profiles if mixing or radionuclide diffusion has not occurred. ^{210}Pb can be used for dating sediments because it is a natural daughter radionuclide in the decay series of ^{238}U. The decay of ^{226}Ra (half-life 1600 years) in soils and sediments produces the rare gas ^{222}Rn (half-life of 3.8 days) which partially diffuses into the atmosphere or into the water column where it decays to ^{210}Pb (half-life of approximately 22 years). ^{210}Pb becomes absorbed onto particles and finally deposits in the bottom sediments (Allen et al., 1993). The ^{210}Pb method is very useful for dating events that have occurred over the last 100-150 years. It has been successfully applied in the sediments of the Ría de Vigo by Álvarez-Iglesias et al. (2007) to obtain sedimentation rates of about 5 mm yr^{-1} in intertidal sediments, whereas Rubio et al. (2001) determined values between 1 and 3 mm yr^{-1} for sediments in inner areas of the Ría de Pontevedra. The analysis of dated sediment cores is tremendously useful because it provides a historical record of natural background levels while it also records the anthropogenic accumulation of metals over the last century.

5.3 The assessment of metal pollution

In order to assess metal pollution in sediment cores it is essential to account for grain-size effects first. The two basic procedures for this purpose are to make analytical determinations on a separate grain-size fraction (Ackerman et al., 1983), or use a normalizing factor to correct the results so that regardless of the sediment size distribution, the analytical results can be compared.

Some authors disagree with the grain-size separation because they think that some metals are associated with the coarser fractions, either as aggregates or pellets composed of fine-grained particles and organic matter, or as grain coatings, that may contain high concentrations of metals. For instance, Rubio et al. (1999) have confirmed the occurrence of pellets and coatings enriched in metals in the sediments of the Rías Baixas. On the contrary, other authors (Araujo et al., 1988; Salomons & Forstner 1984) recommend the use of the fraction smaller than 63 μm in order to minimize grain-size biases on the results of heavy metal content. However, Rubio et al. (1996) concluded that the analysis of this fraction could not compensate for the grain size effect in sediments of the Ría de Pontevedra. For this reason it is always recommended to normalize the metal content to a conservative element such as Al or Rb. In the case of the Galician Rías, the best results for sediments have been obtained with aluminium (Nombela et al. 1994; Marcet Miramontes et al., 1997, Rubio et al., 2000a, 2001). An example for a sediment core from San Simón Bay (inner Ría de Vigo) is shown in figure 4. The similarity of the profiles of the absolute metal concentrations and the Al-normalized results confirm that the increase of Cu, Pb and Zn concentration towards the top of the core is not due to textural effects but anthropogenic inputs. Therefore, the effectiveness of standardization is itself a way to detect metal contamination.

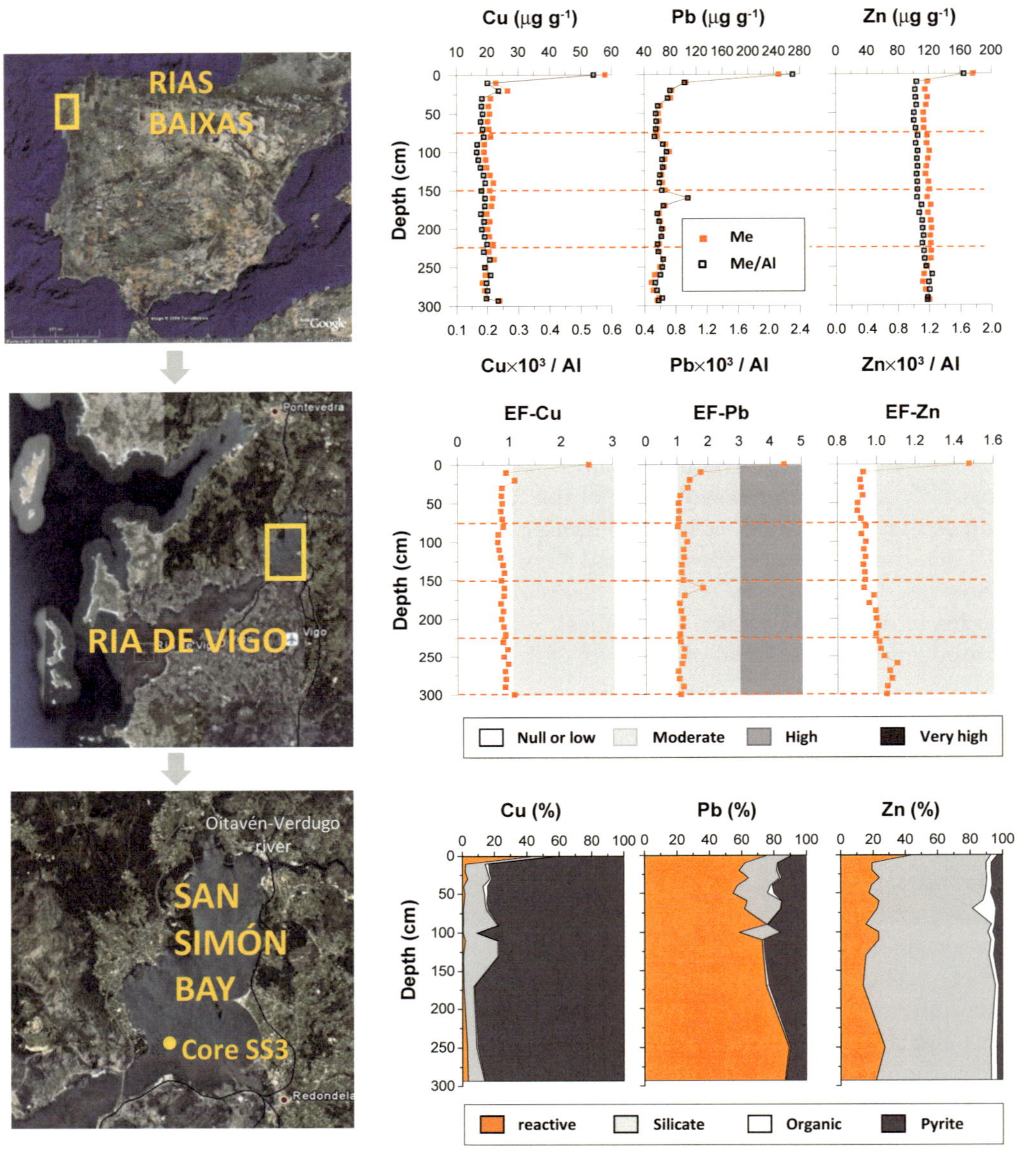

Fig. 4. Left: Location of the core SS3 in the inner area of Ría de Vigo (San Simón Bay). Right: Top, concentrations of Cu, Pb and Zn (orange; data from Álvarez-Iglesias et al., 2006) in a core from San Simón Bay and corresponding metal/Al ratios (black). Middle, enrichment factors (EF) for the same elements and classification of the level of contamination. Bottom, depth distribution of reactive, organic, silicate and pyrite fractions of Cu, Pb and Zn obtained from sequential extractions according to the procedure of Huerta-Díaz & Morse (1990).

Normalized enrichment factors (EF) (Zoller et al., 1974) are also a useful tool where EF = $(M/Al)_{sample}/(M/Al)_{background}$. EFs for Zn, Pb and Cu in core SS3 are shown in figure 4. These results show high contamination for Pb, and moderate for Cu and Zn in the upper part of the core, whereas for the core bottom contamination is moderate for Pb and absent, for Cu and Zn.

5.4 The need to carry out sequential extractions

The total amount of metals in the sediment is unrepresentative of the potential toxicity of the metal. To assess the toxicity appropriately it is essential to know the chemical forms in which a metal is presented, i.e. speciation. The chemical form (as dissolved, adsorbed, bound or precipitated) of an element will not only regulate its degree of toxicity but also its availability.

Total concentrations are still used very frequently in studies of contamination due to its easy measurement and reproducibility, in spite of the fact that the type of contaminant and the form in which it appears (soluble, exchangeable, bound, adsorbed, occluded, etc.) will decisively influence the pollution effect. For this reason, sequential extractions are usually performed and several operationally defined fractions obtained, which depend on the ability of the chemical extractant to remove certain components. These extractions allow us to determine the chemical forms in which each element is found in the sediment. However, very few studies on trace metals in the Galician Rías have considered the forms adopted by different metals. One example of a sequential extraction following the method of Huerta-Díaz & Morse (1990) is shown in figure 4 for inner Ría de Vigo sediments (core SS3). Here we distinguish operationally defined reactive, organic, pyrite and silicate-bound fractions for several trace elements. Pb appeared mostly in the reactive fraction (average, 68.5%), Cu in the pyrite fraction (81.0% on average) and Zn in the silicate-bound fraction (68.7% on average), being the organic fraction very low in all cases (usually lower than 4%). In terms of toxicity these results show that the most problematic trace elements are Pb, because it is found in more biovailable forms, and Cu, because it is found in oxidizable forms. Zn toxicity will mostly come from its reactive fraction. These detailed interpretations confirms the interest of the determination of chemical forms when contamination is suspected in a target area.

5.5 The magnetic properties as a proxy for trace metals in sediments

The measurement of trace elements in sediments is very laborious and expensive and, therefore, the use of fast and economic alternative techniques is desirable. Environmental magnetism -the use of magnetic properties for environmental applications- can be used to estimate contamination levels and assess possible patterns of dispersion of contaminants. Some authors have shown that certain magnetic properties such as magnetic susceptibility (χ) or the isothermal remanent magnetization (IRM) show significant positive correlations with the concentrations of trace metals in the fine-grained fraction of sediments (Chan et al., 1998, 2001; Scoullos & Oldfield, 1984; Spassov et al., 2004), whereas other researchers (Petrovsky et al., 1998) have reported the contrary. In some studies both behaviours are observed depending on the element considered (Berry & Plater 1998; Georgeaud et al., 1997). A positive association is explained by these authors in terms of the preferent absorption of the metals by the clay fraction and Fe oxides, whereas a negative correlation is sometimes explained in terms of diversity of sources of contamination or due to diagenetical processes.

Some studies in the Rías Baixas pointed out that the distribution of magnetic susceptibility in surficial sediments could be explained mostly by the textural and hydrodynamic interplay (Rey et al., 2000, 2005; Mohamed et al., 2011). The increase in diamagnetic carbonate content toward the ría margins, where coarse-grained material accumulates, results in generally low susceptibility values. The highest susceptibilities lie along the central axis, where the clay content is high and carbonate bioclasts are scarce; and also toward the outer sector of the ría, where oxygenation is more intense and formation of authigenic Fe oxides and oxyhydroxides is favoured. The analysis of the susceptibility of the mud fraction (χ_{mud}) that was correlated with trace metals and other properties of the sediments (Fig. 5) revealed a strong negative correlation of susceptibility with Pb. The organic matter content is also correlated with the distribution of elements like Pb, as it is shown in figure 3. Magnetic susceptibility gradually decreases toward the inner part of the central axis because the organic matter decomposition causes reducing conditions and the establishment of an anoxic/sulphidic environment where the magnetic oxides and oxyhydroxides that carry out the susceptibility signal in the outer part of the ría are dissolved. Therefore, low magnetic susceptibility values in sediments of the rías can be a good indicator of reducing conditions, related to organic-rich fine-grained sediments in low-energy environments where trace metals tend to accumulate.

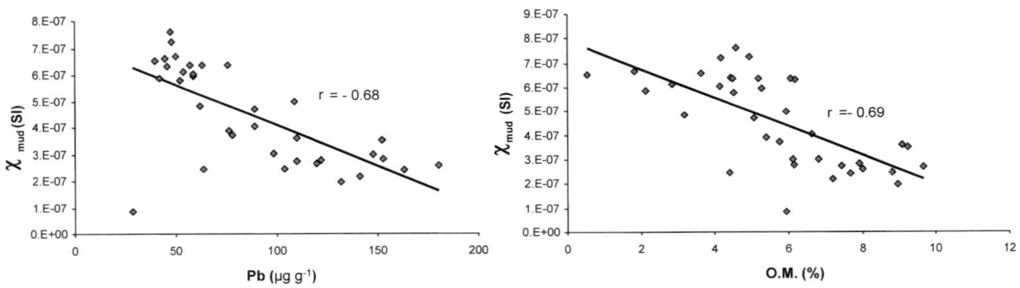

Fig. 5. Relationships between magnetic suspectibility of the mud fraction (χ_{mud}) vs Pb concentration and vs organic matter (O.M.) in surficial sediments from the Rías de Pontevedra and Vigo (modified from López-Rodríguez et al., 1999).

6. Early diagenesis of metals in ría sediments

The early diagenetic reactions that control the formation of authigenic minerals are driven by the oxidation of organic matter, initially by aerobic respiration and subsequently by a series of reactions controlled by anaerobic bacteria, such as reduction of Fe and Mn oxides, reduction of nitrates and sulfates and methanogenesis (Canfield et al., 1993). These reactions release products (e.g., HCO_3^-, HS^-, Fe^{+2}, Mn^{+2}) to the sediment pore waters, which will precipitate forming new minerals when the saturation is reached (Gaillard et al., 1989). These processes occur ideally sequentially starting with oxic, suboxic, sulfidic and finally methanogenic reactions (Berner, 1981). This diagenetic sequence of events can be evaluated from the analysis of pore-waters and the mineral concentration of typical diagenetic mineral phases in sediment cores, or by using sequential extractions in the sediments.

6.1 Diagenetic zonation in ría environments: The hydrodynamic role

Previous studies in the Rías Baixas allowed the definition of a diagenetic zonation model in these environments by using a combination of geochemical sequential extractions and magnetic properties (Mohamed et al., 2011; Rey et al., 2005; Rubio et al., 2001, 2010). In particular, speciation data of redox sensitive elements such as Fe and Mn are indicative of the different reducing conditions in sediments. Magnetic properties are useful to identify the magnetic minerals and their concentration, which can be used as proxies for the different diagenetic environments. Figure 6 shows the deepening of the redox boundaries from inner to outer ría. The oxic zone expands as it gets deeper toward the outer ría, in a similar way as the suboxic, anoxic and methanic zones. The observed redoxcline deepening can also be related to the different depths at which shallow gas fields have been described in the Ría de Vigo sediments (García-Gil et al., 2002; Kitidis et al., 2006; Iglesias and García-Gil, 2007).

This spatial trend can be explained by several factors: 1) A progressive change in the hydrodynamic conditions along the ría, 2) The different origin (marine or terrestrial) of the organic matter and their aging in the water column.

Regarding hydrodynamics the outermost ría areas are affected by severe storms in winter that remobilize and oxygenate the top sediments due to wave action. This process buffers sulphate reduction and contributes to the formation of authigenic iron oxides by precipitation of dissolved iron diffusing from underlying anoxic layers (Rubio et al., 2001). This process also seems to contribute to the gradual depletion of organic matter in fine-grained sediments observed toward the outer areas of the ría mouth.

As for the organic matter characteristics, the decrease in terrestrial sedimentary organic matter toward the outer-ría (Álvarez-Iglesias 2006; Andrade et al. 2011), in addition to the longer aging of organic matter in deeper waters of the outer-ría areas compared to the inner-ría, could contribute to explain the mentioned diagenetic zonation.

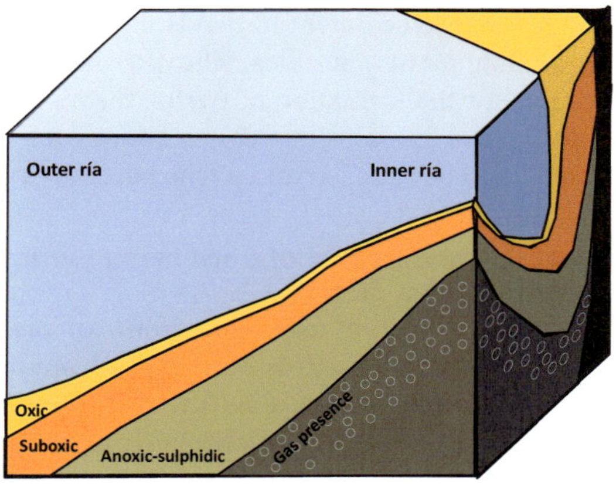

Fig. 6. Block diagram illustrating the variation in depth of the diagenetic zones in the different sectors of the ría.

6.2 Diagenetic mobilization of trace metals: Influence of mussel rafts

In the last fifty years mussel culture in the Rías Baixas has caused significant changes in the sediment due to the large amounts of detritus originated by these filter feeders, which are deposited mostly as pellets enriched in organic matter on the ría bottoms. Each mussel raft produces approximately 190 kg day^{-1} of dry biodeposit that contains between 31 and 32 kg day^{-1} of organic matter (Cabanas et al., 1979). In addition, sedimentation rate is increased significantly in areas below mussel rafts, with values that range between 5 and 15 mm yr^{-1} (Tenore & González, 1975; Cabanas et al., 1979). This elevated sediment accumulation together with its high concentration of organic matter has led to a change in the physico-chemical properties of sediments towards more anoxic environments (Cabanas et al., 1982; León et al., 2004).

In these sediments, anoxic degradation of organic matter is responsible for the early diagenesis of sedimentary Fe sulfides that eventually are transformed into pyrite (FeS_2), which is thermodynamically the more stable compound (Luther, 1991; Morse & Luther, 1999).

The study of diagenesis and organic matter degradation can provide very important information about retention and/or mobility of contaminants such as trace metals. Some authors considered that formation of insoluble sulfides under reducing conditions would immobilize and trap trace metals such as Cu, Zn and Pb. On the contrary, other studies (Álvarez-Iglesias & Rubio, 2008, 2009; Rae & Allen, 1993; Rubio et al., 2010; Varekamp, 1991) indicated that these elements can be mobilized or relocated during the degradation of organic matter. It is also important to distinguish between the fraction of the elements incorporated in detrital phases and the fraction which may be available in response to changes in redox conditions, such as variations in the chemical conditions of the bottom water or interstitial water. The two main approaches to make this separation are: 1) The use of statistical techniques of separation of these phases (Calvert, 1976; Dymond, 1981). 2) The application of chemical treatments to remove certain phases or fractions of elements (Huerta- Díaz & Morse, 1990; Tessier et al., 1979; Ure et al., 1993, among others). As we have seen in section 5.4 the latter approach, sequential extractions, are a key tool to assessing the bioavailability of a particular metal.

The availability of trace metals in the sediment depends on the fractions to which they are associated to (carbonates, organic matter, sulfides, silicates, oxyhydroxides of Fe and Mn). When conditions are favorable for the formation of pyrite, metals can co-precipitate with it, and pyrite becomes an important metal sink. If environmental conditions change (i.e. oxidation of sediments) metals can be released and pyrite becomes a source.

6.3 Degree of pyritization (DOP) and degree of trace metal pyritization (DTMP) as proxies for predicting mobilization of metals

Two parameters (DOP and DTMP) can be used as proxies for predicting mobilization of metals. The DOP has been used to classify sedimentary marine environments because it is a useful paleoenvironmental geochemical index that has been correlated and corroborated with paleoecological data (Raiswell et al., 1988). The DOP is calculated from the reactive fractions (extracted with HCl) and pyrite (Berner, 1970) as $DOP = [Fe_{pyr}/ (Fe_{react} + Fe_{pyr})]*100$. Similarly we can determine DTMP according to Huerta-Díaz & Morse (1990) as$DTMP = [M_{pyr}/(M_{react} + M_{pyr})]*100$, where M is the metal of interest.

In order to homogenize the differences in nomenclatures, León et al. (2004) proposed a new classification that combines the Berner's (1981) pioneering geochemical classification of sedimentary environments and the above mentioned work of Raiswell et al. (1988) based on the DOP. This new classification establishes that the sedimentary environments are oxic when DOP is <42%, dioxic or suboxic (42% - 55%), anoxic (55%- 75%), and euxinic (DOP >75%).

Based on this classification we compared the DOP for sediment cores in inner and middle sectors of the Ría de Pontevedra (Fig. 7). The cores of the middle sector, on average, have lower DOP values and fall into the oxic category, whilst the redox status of the cores of the inner sector, vary from suboxic to anoxic. The presence of anoxic areas in the rías is an unusual situation in an ecosystem where the dissolved oxygen in the water column is not completely exhausted, in spite of the high biological productivity in these areas resulting from the upwelling process (Figueiras et al., 1986). However, due to the high sedimentation rate of organic-rich material, especially under mussel rafts, anoxic environments can be developed at the sediment-water interface. Regarding the DTMP (Fig. 7), and its differences between mussel rafts areas and adjacent sediments the highest values correspond to Hg in both cases. The sources of Hg in the Ría de Pontevedra are paper pulp and electrochemical companies (ELNOSA-ENCE complex), in operation since the 1950's. This is a typical example of point-source contamination, where total Hg concentrations in sediments above 2 $\mu g\ g^{-1}$ (Fig. 8) are detected close to the discharge area of this industrial complex. However, Hg was mainly associated with pyrite phases (Hg_{pyr}), while reactive Hg (Hg_{react}) is only detected in the upper 25 cm of the core and in much lower concentrations than Hg_{pyr}.

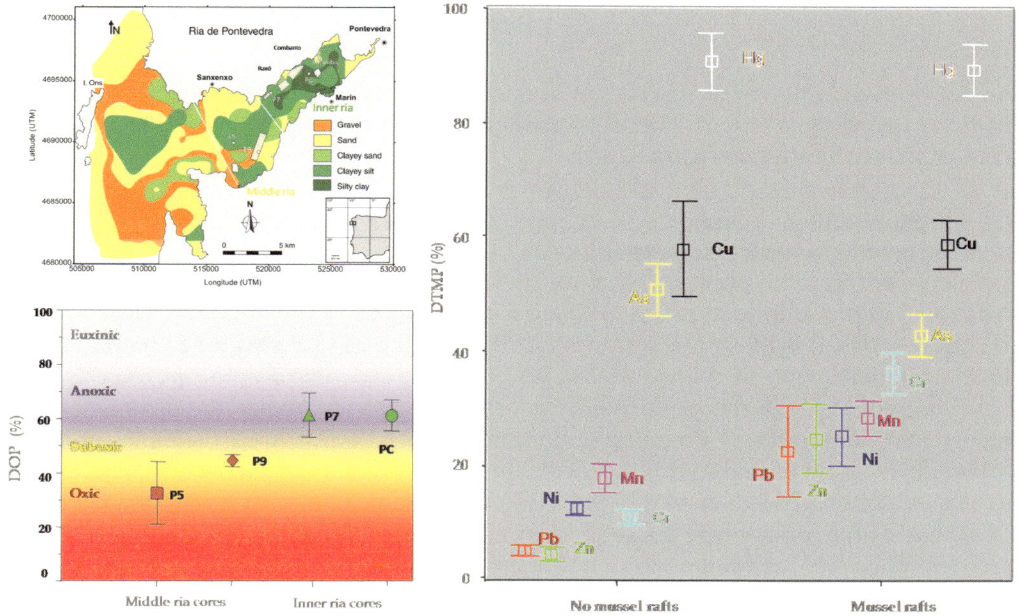

Fig. 7. a) Sediment distribution map of the Ría de Pontevedra (modified from Vilas et al., 1995) and core location. b) Sedimentary environments of the Ría de Pontevedra based on the degree of pyritization (DOP, % mean values) for these cores. The limits for oxic, suboxic, anoxic and euxinic are from León et al. (2004). c) Mean DTMP (%) values for cores both influenced by mussel rafts and from adjacent areas with no mussel rafts.

The Hg_{react} is weakly adsorbed to the components of the sediment matrix and can be released, relatively easily, into the water column due to changes in environmental conditions such as sediment resuspension. In addition, pyrite can be oxidized and release the Hg bound to its structure, constituting a serious threat to aquatic fauna, especially fish and filter feeders. In general terms, the action of water currents, bioturbation, or human activities can provoke remobilization of trace metals associated with organic matter or reduced forms (Otero et al., 2000, 2005; Rubio et al., 2008) that may eventually contaminate the interstitial and suprajacent waters. For a more precise assessment of the water pollution risks that such events can produce, aerobic oxidation experiments such as the one shown in the next section are needed.

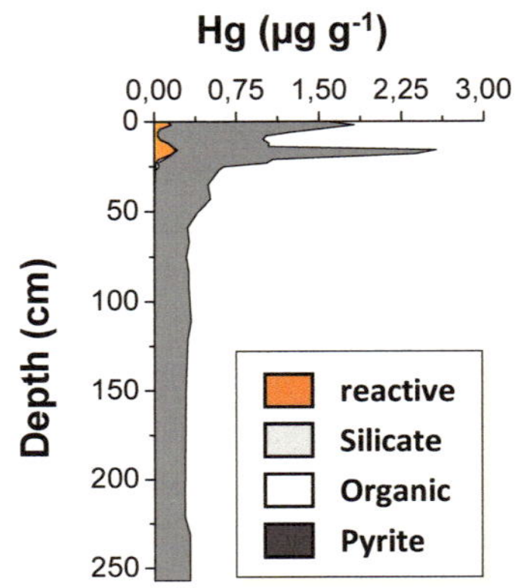

Fig. 8. Depth-distribution of Hg concentrations in reactive, silicate, organic and pyrite fractions according to Huerta-Diaz & Morse (1990) in a core from the Ría de Pontevedra (core PC). Core location in figure 7.

6.4 Solubility of heavy metals during controlled aerobic oxidation of anoxic sediments: Some laboratory experiments

As we have seen in previous sections the concentration levels of certain metal and metalloids in the sediments of the Galician Rías Baixas have shown an increasing trend in the last decades. It is likely that a transfer of these elements to the water column may occur during remobilization of sediments caused by natural events or anthropogenic activities. The inner areas of the rías are exposed to activities that remobilize the sediment such as intense maritime traffic or dredging and cleaning operations. Selected samples of surficial sediments from inner and middle ría sediments of Ría de Pontevedra were subjected to an aerobic oxidation procedure to determine the concentration of some elements (Fe, Mn, Cu, Cr, Pb and Hg) released from the sediment to the aqueous phase. The experiment was done over five days and measurements of pH and total metal concentrations were made both in water and in sediment samples. Metal concentrations were lower in the sediments during aerobic oxidation due to their release to the aqueous phase.

The net release of metals was higher in sediments form the inner sector than those from the middle sector of the Ría de Pontevedra (Fig. 9), with the exception of Cu. The high standard deviation of Fe and Mn in the inner sector samples is mainly due to the high redox sensitivity of these two metals and their high abundance as sulphides, as we have mentioned concering the DOP values, which are rapidly oxidized causing the release of these metals to the aqueous phase. The concentrations of these metals together with those of Cu, Cr and Zn increased significantly in the aqueous phase after the experiment. This demonstrates that remobilization of marine sediments tends to increase the mobility and availability of those trace metals.

Metal concentrations in the aqueous phase varied between elements (Fig. 9). Hg and Pb concentrations were below the detection limits in all cases. Cr and Zn concentrations were in general quite low and remained almost constant over time. In contrast, Fe and Mn were released very rapidly although their concentrations decreased sharply to reach undetectable limits, because they precipitated as oxides and oxyhydroxides. Finally, the release of Cu increased with time for most of the samples, with a maximum concentration of total dissolved Cu of 8.9 mg L^{-1}. This concentration is higher than the toxicity threshold for organisms of the Galician Rías reported by other authors (Beiras & Albentosa, 2004).

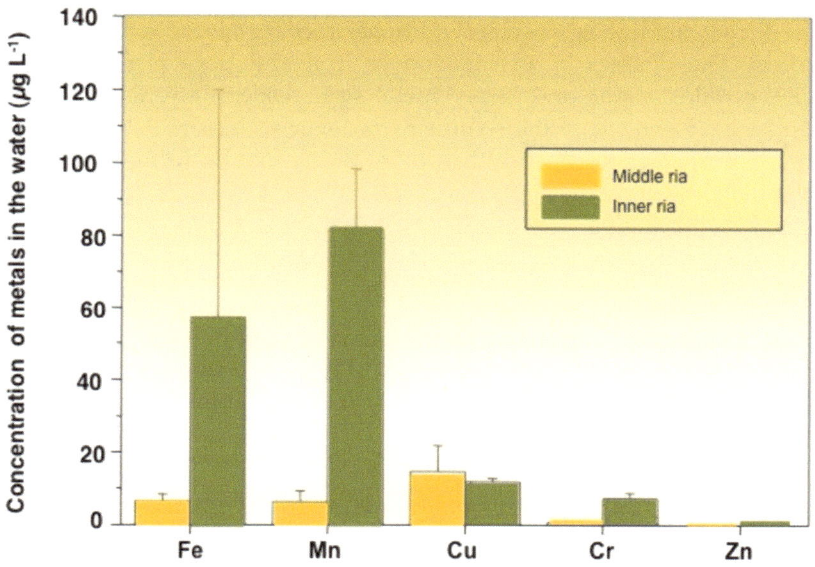

Fig. 9. Mean concentrations of trace metals in water after oxidation of sediments from inner and middle areas of Ría de Pontevedra. Sampling location in figure 7.

7. Conclusion

The main factors controlling the incorporation of metals to the sediments in transitional coastal environments like the Galician Rías Baixas in the NW Spain have been revised in this chapter. It is essential to understand the behaviour of trace metals in the sediments from the ría bottoms in order to improve coastal risks prevention and management, as well as to reach a good status in water quality as one of the great challenges for the European Union in the new millennium.

In the rías, wave conditions exert an important control on sediment distribution and in the subsequent diagenetic evolution of the sediments, and thus on trace metal accumulation and immobilization. A strong positive correlation is found between fine-grained sediments and trace metals. Organic matter enhances the grain-size effect on metal concentration, especially in areas influenced by mussel culture. The procedures for normalizing and minimizing grain-size effects have also been revised in order to distinguish natural from anthropogenic metal signals in the sediments.

Inner ría sediments are highly contaminated by Pb, and moderately by Cu and Zn, especially in the most recent sedimentary record. Some examples of point-source Hg contamination have also been shown for the Ría de Pontevedra. The importance of distinguishing and quantifying the various metal forms by using sequential extractions have also been demonstrated with several examples for sediment cores, highlighting the role of the diagenetic processes in the inmobilization and/or relocation of trace metals. A characteristic diagenetic zonation in ría environments is attributed to the local water depth, the distribution of wave energy and the subsequent sediment grain-size distribution. The diagenetic processes have also been emphasized by the results of the magnetic properties, showing that low magnetic susceptibility values in sediments of the rías can be a good indicator of reducing conditions. In such conditions, trace metals are mostly concentrated in sulfide fractions. The degrees of pyritization of iron and trace elements can be valid indicators of the redox status and heavy metal risk, respectively. Experimental aerobic oxidation results have show that the sediments of inner sectors of the rías show a higher release of metals to the aqueous phase than those of the middle sector. However, from an environmental point of view, Cu is the only metal released in quantities that may be toxic for the organisms in the area.

8. Acknowledgment

This work was supported by the Spanish Ministry of Science and Technology through projects CTM2007-61227/MAR, GCL2010-16688 and IPT-310000-2010-17, by the IUGS-UNESCO through project IGCP-526 and by the Xunta de Galicia through projects 09MMA012312PR and 10MMA312022PR.

9. References

Ackerman, F., Bergmann, H. & Schleichert, U. (1983). Monitoring of heavy metals in coastal and estuarine sediments. A question of grain size: <20 µm versus <60 µm. *Environmental and Technological Letters*, Vol. 4, pp. 317-328.

Allen, J.R.L., Rae, J.E., Longworth, G., Hasler, S.E. & Ivanovich, M. (1993). A comparison of the ^{210}Pb dating technique with three other independent dating methods in an oxic estuarine salt-marsh sequence. *Estuaries and Coasts*, Vol. 16, pp. 670-677.

Álvarez-Iglesias, P. (2006). *El registro sedimentario reciente de la Ensenada de San Simón (Ría de Vigo, Noroeste de España): interacción entre procesos naturales y actividades antropogénicas*. Ph.D. Thesis, Universidad de Vigo, 356 p.

Álvarez-Iglesias, P., Quintana, B., Rubio, B. & Pérez-Arlucea, M. (2007). Sedimentation rates and trace metal input history in intertidal sediments derived from ^{210}Pb and ^{137}Cs chronology. *Journal of Environmental Radioactivity*, Vol. 98, pp. 229-250.

Álvarez-Iglesias, P. & Rubio, B. (2008). The degree of trace metal pyritization in subtidal sediments of a mariculture area: application to the assessment of toxic risk. *Marine Pollution Bulletin*, Vol. 56, pp. 973–983.

Álvarez-Iglesias, P. & Rubio, B. (2009). Redox status and heavy metal risk in intertidal sediments in NW Spain as inferred from the degrees of pyritization of iron and trace elements. *Marine Pollution Bulletin*, Vol. 58, pp. 542-551.

Álvarez-Iglesias, P., Rubio, B. & Pérez-Arlucea, M. (2006). Reliability of subtidal sediments as "geochemical recorders" of pollution input: San Simón Bay (Ría de Vigo, NW Spain). *Estuarine, Coastal and Shelf Science,* Vol. 70, pp. 507-521.

Álvarez-Iglesias, P., Rubio, B. & Vilas, F. (2003). Pollution in intertidal sediments of San Simón Bay (Inner Ría de Vigo, NW of Spain): total heavy metal concentrations and speciation. *Marine Pollution Bulletin,* Vol. 46, pp. 491-506.

Andrade, A., Rubio, B., Rey, D., Álvarez-Iglesias, P., Bernabeu, A.M. & Vilas, F. (2011). Paleoclimatic changes in the northwestern Iberia península during the last 3000 years as inferred from diagenetic proxies in the sedimentary record of the Ría de Muros. *Climate Research,* Vol. 48, pp. 247-259

Angelidis, M.O. & Aloupi, M. (1995). Metals in sediments of Rhodes Harbour, Greece. *Marine Pollution Bulletin,* Vol. 31, pp. 273-276.

Araujo, M.F., Bernard, P. C. & Van Grieken, R. E. (1988). Heavy metal contamination in sediments from the Belgian Coast and Sheldt estuary. *Marine Pollution Bulletin,* Vol. 19, pp. 269-273.

Barreiro Lozano, R., Carballeira Ocaña, A. & Real Rodríguez, C. (1988). Metales pesados en los sedimentos de cinco sistemas de ría (Ferrol, Burgo, Arousa, Pontevedra y Vigo). *Thalassas,* Vol. 6, pp. 61-70.

Barreiro, R. (1991). Estudio de metales pesados en medio y organismos de un ecosistema de ría (Pontedeume, A Coruña). Ph.D. Thesis, Universidad de Santiago de Compostela, 227 p.

Beiras, R. & Albentosa, M. (2004). Inhibition of embryo development of the commercial bivalves *Ruditapes decussatus* and *Mytilus galloprovincialis* by trace metals; implications for the implementation of seawater quality criteria. *Aquaculture,* Vol. 230, pp. 205-213.

Belzunce Segarra, M.J., Prego, R., Wilson, M.J., Bacon, J. & Santos-Echeandía, J. (2008). Metal speciation in surface sediments of the Vigo Ria (NW Iberian Peninsula). *Scientia Marina,* Vol. 72, pp. 119-126.

Berner, R.A. (1970). Sedimentary pyrite formation. *American Journal of Science,* Vol. 268, pp. 1-23.

Berner, R. A. (1981). A new geochemical classification of sedimentary environments. *Journal of Sedimentary Petrology,* Vol. 51, No.2, pp. 359-365.

Berry, A. & Plater, J. (1998). Rates of tidal sedimentation from records of industrial pollution and environmental magnetism: the Tees estuary, North-East England. *Water, Air and Soil Pollution,* Vol. 106, pp. 463-478.

Cabanas, J., González, J. & Iglesias, M. (1982). Physico-chemical conditions in winter in the Ría de Pontevedra (NW Spain) and their influences on contamination. *International Council for the Exploration of the Sea-Conseil International pour l'Exploration de la Mer* (ICES CIEM), Vol. E53, pp. 15.

Cabanas, J., Mariño, J., Pérez, A. & Román, G. (1979). Estudio del mejillón y de su epifauna en los cultivos flotantes de la ría de Arousa. III. Observaciones previas sobre la retención de partículas y la biodeposición de una batea. *Boletin del Instituto Español de Oceanografía,* Vol. 268, pp. 45-50.

Calvert, S.E. (1976). The mineralogy and geochemistry of nearshore sediments, In: *Chemical Oceanography,* J.P. Riley & R. Chester (Eds), vol. 6, 187-280, Academic Press, London.

Calvo de Anta, R., Quintas Mosteiro, Y. & Macías Vázquez, F. (1999). Caracterización de materiales para la recuperación de suelos degradados. I: Sedimentos biogénicos de las Rías de Galicia. *Edafología*, Vol. 6, pp. 47-58.

Canfield, D. E. Thamdrup, B. & Hansen, J. W. (1993). The anaerobic degradation of organic matter in Danish coastal sediments: iron reduction, manganese reduction and sulfate reduction. *Geochimica et Cosmochimica Acta*, Vol. 57, pp. 3867-3883.

Carral, E., Villares, R., Puente, X. & Carballeira, A. (1995). Influence of watershed lithology on heavy metal levels in estuarine sediments and organisms in Galicia (North-west Spain). *Marine Pollution Bulletin*, Vol. 30, pp. 604-608.

Chan, L.S., Ng, S.L., Davis, A.M., Yim, W.W.S. & Yeung, C.H. (2001). Magnetic properties and heavy-metal contents of contaminated seabed sediments of Penny's Bay, Hong Kong. *Marine Pollution Bulletin*, Vol. 42, pp. 569-583.

Chan, L.S., Yeung, C.H., Yim, W.S.-W. & Or, O.L. (1998). Correlation between magnetic susceptibility and distribution of heavy metals in contaminated sea-floor sediments of Hong Kong Harbour. *Environmental Geology*, Vol. 36, pp. 77-86.

Chester, R. & Voutsinou, F. G. (1981). The initial assessment of trace metal pollution in coastal sediments. *Marine Pollution Bulletin*, Vol. 12, pp. 84-91.

Dymond, J. (1981). Geochemistry of Nazca plate surface sediments: An evaluation of hydrothermal, biogenic, detrital and hydrogenous sources. *Geological Society of America Memoir*, Vol. 154, pp. 133-172.

EC (European Communities), 2000. Directive 2000/60/EC of the European Parliament and of the Council of 23 October 2000 establishing a framework for Community action in the field of water policy. *Official Journal of the European Communities*, L327/1, 22 December 2000.

ECJ (European Court of Justice), 2005. Case C-26/04 ECJ, Commission v Spain ECJ 15- 12-2005, ECRI-11059.

EEC (European Economic Community), 1979. Council Directive 79/923/EEC of 30 October 1979 on the Quality required of Shellfish Waters. *Official Publications of the European Communities*, OJ L281, 10 November 1979.

Figueiras, F., Niell, F. & Mouriño, C. (1986). Nutrientes y oxígeno en la Ría de Pontevedra (NO de España). *Investigaciones Pesqueras*, Vol. 50, pp. 97-115.

Förstner, U. & Wittmann, G. T. (1981). *Metal pollution in the aquatic environment*. Springer-Verlag, London, 486 p.

Fraga, F. (1981). Upwelling off the Galician Coast, NE Spain. In: *Coastal Upwelling, Coastal Estuarine Studies*, F.A. Richards (Ed.), 1, 176-182, American Geophysical Union, Washington, DC.

Gaillard, J.F., Pauwells, H. & Michard, G. (1989). Chemical diagenesis in coastal marine sediments. *Oceanologica Acta*, Vol. 12, No.3, pp. 175-187.

García-Gil, S., Vilas, F. & García-García, A. (2002). Shallow gas features in incised-valley fills (Ria de Vigo, NW Spain): a case study. *Continental Shelf Research*, Vol. 22, pp. 2303-2315.

Georgeaud, V.M., Rochette, P., Ambrosi, J.P., Vandamme, D. & Williamson, D. (1997). Relationship between heavy metals and magnetic properties in a large polluted catchment: The Etang de Berre (South of France). *Physics and Chemistry of the Earth*, vol. 22, pp. 211-214.

Horowitz, A.J. & Elrick, K.A. (1987). The relation of stream sediment surface area, grain size, and composition of trace element chemistry. *Applied Geochemistry*, Vol. 2, pp. 437-451.

Huerta-Díaz, M.A. & Morse, J. (1990). A quantitative method for determination of trace metal concentrations in sedimentary pyrite. *Marine Chemistry*, Vol. 29, pp. 119-144.

Iglesias, J. & García-Gil. S. (2007). High-resolution mapping of shallow gas accumulations and gas seeps in San Simón Bay (Ría de Vigo, NW Spain). *Geo-Marine Letters*, Vol. 27, pp. 103-114.

Kitidis, V., Tizzard, L., Uher, G., Judd, A.G., Upstill-Goddard, R., Head, I.M., Gray, N.D., Taylor, G., Durán, R., Diez, R., Iglesias, J. & García-Gil, S. (2006). The biogeochemical cycling of methane in Ría de Vigo, NW Spain. *Journal of Marine Systems*, Vol. 66, pp. 258-271.

Lee, S.V. & Cundy, A.B. (2001). Heavy metal contamination and mixing processes in sediments from the Humber Estuary, Eastern England. *Estuarine, Coastal and Shelf Science*, Vol. 53, pp. 619-636.

Lei 8/2001, of August 2nd, de protección da calidade das augas das rías de Galicia e de ordenación do servicio público de depuración de augas residuais urbanas. *Diario Oficial de Galicia* (DOG).

León, I., Méndez, G. & Rubio, B. (2004). Geochemical phases of Fe and degree of pyritization in sediments from Ría de Pontevedra (NW Spain): Implications of mussel raft culture. *Ciencias Marinas*, Vol. 30, pp. 585-602.

Ligero, R.A., Barrera, M., Casas-Ruiz, M., Sales, D. & López-Aguayo, F. (2002). Dating of marine sediments and time evolution of heavy metal concentrations in the Bay of Cádiz, Spain. *Environmental Pollution*, Vol. 118, pp. 97-108.

Long, E.R., MacDonald, D.D., Smith, S.L. & Calder, F.D. (1995). Incidence of adverse biological effects within ranges of chemical concentrations in marine and estuarine sediments. *Environmental Management*, Vol. 19, pp. 81-97.

López-Rodríguez, N., Rey, D., Rubio, B., Pazos, O. & Vilas, F. (1999). Variaciones de la susceptibilidad magnética en los sedimentos de la Ría de Vigo (Galicia). Implicaciones para la dinámica sedimentaria y contaminación antropogénica de la zona. *Thalassas*, Vol. 15, pp. 85-94.

Luther, G.W. III. (1991). Pyrite synthesis via polysulfide compounds. *Geochimica et Cosmochimica Acta*, Vol. 55, pp. 2839-2849.

Marcet Miramontes, P., Andrade Couce, M. L., & Montero Vilariño, M. J. (1997). Contenido y enriquecimiento de metales en sedimentos de la Ría de Vigo (España). *Thalassas*, Vol. 13, pp. 87-97.

Mohamed, K.J., Rey, D., Rubio, B., Dekkers, M., Roberts, A.P. & Vilas, F. (2011). Onshore-offshore gradient in reductive early diagenesis in coastal marine sediments of the Ría de Vigo, Northwest Iberian Peninsula. *Continental Shelf Research*, Vol. 31, No.5, pp. 433-447.

Morse, J. W. & Luther, G. W., III (1999). Chemical influence on trace metal-sulfide interactions in anoxic sediments. *Geochimica et Cosmochimica Acta*, Vol. 63, No.19/20, pp. 3373-3378.

Mucha, A.P., Vasconcelos, M.T.S.D. & Bordalo, A.A. (2003). Macrobenthic community in the Douro estuary: relations with trace metals and natural sediment characteristics. *Environmental Pollution*, Vol. 121, pp. 169-180.

Nombela, M. A., Vilas, F., García- Gil, S., García- Gil, E., Alejo, I., Rubio, B. & Pazos, O. (1994). Metales pesados en el registro sedimentario reciente en la Ensenada de San Simón, parte interna de la Ría de Vigo (Galicia, España). *Gaia*, Vol. 8, pp. 149-156.

Otero, X.L., Calvo de Anta, R.M. & Macías, F. (2006). Sulphur partitioning in sediments and biodeposits below mussel rafts in the Ría de Arousa (Galicia, NW Spain). *Marine Environmental Research*, Vol. 61, pp. 305-325.

Otero, X. L., Vidal-Torrado, P., Calvo de Anta, R. M. & Macías, F. (2005). Trace elements in biodeposits and sediments from mussel culture in the Ría de Arousa (Galicia, NW Spain). *Environmental Pollution*, Vol. 136, pp. 119-134.

Otero, X.L., Sánchez, J.M. & Macías, F. (2000). Bioaccumulation of heavy metals in thionic fluvisols by a marine polychaete: the role of metal sulfides. *Journal of Environmental Quality*, Vol. 29, pp. 1133–1141.

Pekey, H., Karakaş, D., Ayberk, S., Tolun, L., & Bakoğlu, M. (2004). Ecological risk assessment using trace elements from surface sediments of İzmit Bay (Northeastern Marmara Sea) Turkey. *Marine Pollution Bulletin*, Vol. 48, pp. 946–953.

Petrovsky, E., Kapicka, A., Zapletal, K., Sebestova, E., Spanila, T., Dekkers, M.J., & Rochette, P. (1998). Correlation between magnetic parameters and chemical composition of lake sediments from northern Bohemia-preliminary study. *Physics and Chemistry of the Earth*, Vol. 23, pp. 1123-1126.

Piedracoba, S., Souto, C., Gilcoto, M. & Pardo, P.C. (2005). Hydrography and dynamics of the Ría de Ribadeo (NW Spain), a wave driven estuary. *Estuarine Coastal and Shelf Science*, Vol. 65, pp. 726-738.

Prego, R. & Cobelo-Garcia, A. (2003). Twentieth century overview of heavy metals in the Galician Rias (NW Iberian Peninsula). *Environmental Pollution*, Vol. 121, pp. 425–452.

Prego, R., Otxotorena, U. & Cobelo-García, A. (2006) Presence of Cr, Cu, Fe and Pb in sediments underlying mussel-culture rafts (Arosa and Vigo rias, NW Spain). Are they metal-contaminated areas? *Ciencias Marinas*, Vol. 32, No.2B, pp. 339-349.

Rae, J.E. & Allen, J.R.L. (1993). The significance of organic matter degradation in the interpretation of historical pollution trends in depth profiles of estuarine sediment. *Estuaries*, Vol. 16, No.3B, pp. 678-682.

Raiswell, R. & Berner, R.A. (1986). Pyrite and organic matter in Phanerozoic normal marine shales. *Geochimica et Cosmochimica Acta*, Vol. 50, pp. 1967-1976.

Raiswell, R., Buckley, F., Berner, R. & Anderson, T. (1988). Degree of pyritization of iron as a paleoenvironmental indicator of bottom-water oxygenation. *Journal of Sedimentary Petrology*, Vol. 58, pp. 812-819.

Rey, D., López-Rodríguez, N., Rubio, B., Vilas, F., Mohamed, K., Pazos, O. & Bógalo, M.F. (2000). Magnetic properties of estuarine-like sediments. The study case of the Galician Rías. *Journal of Iberian Geology*, Vol. 26, pp. 151-170.

Rey, D., Mohamed, K., Bernabeu, A., Rubio, B. & Vilas, F. (2005). Early diagenesis of magnetic minerals in marine transitional environments: geochemical signatures of hydrodynamic forcing. *Marine Geology*, Vol. 215, pp. 215–236.

Rubio, B., Álvarez-Iglesias, P. & Vilas, F. (2010). Diagenesis and anthropogenesis of metals in the recent Holocene sedimentary record of the Ría de Vigo (NW Spain). *Marine Pollution Bulletin*, Vol. 60, pp. 1122-1129.

Rubio, B., León, I., Álvarez-Iglesias, P. & Vilas, F. (2008). Aerobic oxidation of suboxic-anoxic sediments: implications for metal remobilization and release. *Geotemas*, Vol. 10, pp. 651-654.

Rubio, B., Gago, L., Vilas, F., Nombela, M.A., García-Gil, S., Alejo, I. & Pazos, O. (1996). Interpretación de tendencias históricas de contaminación por metales pesados en testigos de sedimentos de la Ría de Pontevedra. *Thalassas*, Vol. 12, pp. 137-152.

Rubio, B., Nombela M.A. & Vilas F. (2000a). Geochemistry of major and trace elements in sediments of the Ría de Vigo (NW Spain): An assessment of metal pollution. *Marine Pollution Bulletin*, Vol. 40, pp. 968-980.

Rubio, B., Nombela M.A. & Vilas F. (2000b). La contaminación por metales pesados en las Rías Baixas gallegas: nuevos valores de fondo para la Ría de Vigo (NO de España). *Journal of Iberian Geology*, Vol. 26, pp. 121-149.

Rubio, B., Pye, K., Rae, J. & Rey, D. (2001). Sedimentological characteristics, heavy metal distribution and magnetic properties in subtidal sediments, Ría de Pontevedra, NW Spain. *Sedimentology*, Vol. 48 No.6, pp. 1277-1296.

Rubio, B., Rey, D., Pye, K., Nombela, M.A. & Vilas, F. (1999). Aplicación de imágenes de electrones retrodispersados en microscopía electrónica de barrido a sedimentos litorales. *Thalassas*, Vol. 15, pp. 71-84.

Ruiz-Villarreal, M., Montero, P., Taboada, J.J., Prego, R., Leitão, P.C. & Pérez-Villar, V. (2002). Hydrodynamic model study of the Ria de Pontevedra under estuarine conditions. *Estuarine Coastal and Shelf Science*, Vol. 54, pp. 101-113.

Salomons, W. & Förstner, U. (1984). *Metals in the hydrocycle*. Springer-Verlag, Berlin, 349 p.

Scoullos, M. & Oldfield, F. (1984). Mineral magnetic studies for a pollution monitoring of marine and estuarine sediments. *VIIes Journal Etudes Pollution*, CIESM/COI/PNUE, Lucerne.

Soto-Jiménez, M.F., Arellano-Fiore, C., Rocha-Velarde, R., Jara-Marini, M.E., Ruelas-Inzunza, J. & Páez-Osuna, F. (2011). Trophic transfer of lead through a model marine four-level food chain: *Tetraselmis suecica, Artemia franciscana, Litopenaeus vannamei*, and *Haemulon scudderi*. *Archives of Environmental Contamination and Toxicology*, Vol. 61, No2, pp. 280-291

Souto, C., Gilcoto, M., Fariña-Busto, L. & Pérez, F.F. (2003). Modeling the residual circulation of a coastal embayment affected by wind-driven upwelling: Circulation of the Ría de Vigo (NW Spain). *Journal of Geophysical Research*, Vol. 108, No.C11, pp. 3340-3356.

Spassov, S., Egli, R., Heller, F., Nourgaliev, D.K. & Hannam, J. (2004). Magnetic quantification of urban pollution sources in atmospheric particulate matter. *Geophysical Journal International*, Vol. 159, pp. 555-564.

Subramanian, V. & Mohanachandran, G. (1990). Heavy metals distribution and enrichment in the sediments of Sourthern East Coast of India. *Marine Pollution Bulletin*, Vol. 21, pp. 324-330.

Summers, J.K., Wade, T.L., & Engle, V.D. (1996). Normalization of metal concentrations in estuarine sediments from the Gulf of Mexico. *Estuaries*, Vol. 19, No.3, pp. 581-594.

Taylor, S.R. (1964). Abundance of chemical elements in the continental crust: a new table. *Geochimica et Cosmochimica Acta*, Vol. 28, pp. 1273-1285.

Tenore, K. & González, N. (1975). Food chain patterns in the Ría de Arosa, Spain: An area of intense mussel aquaculture, In: *10th European Symposium of Marine Biology*, G. Persoone & E. Jaspers (Eds.), Vol. 2, pp. 601-619, Ostend, Belgium, September 17-23, 1975, Universa Press, Wetteren.

Tessier, A. & Campbell, P.G.C. (1988). Partitioning of trace metals in sediments, In: *Metal Speciation: Theory, Analysis and Application*, J.R. Kramer & H.E. Allen (Eds.), 183-199, Lewis Publishers, Inc.

Tessier, A., Campbell P.G.C. & Bisson, M. (1979). Sequential extraction procedure for the speciation of particulate trace metals. *Analytical Chemistry*, Vol. 51, pp. 844-851.

Turekian, K.K. & Wedepohl, K.H. (1961). Distribution of the elements in some major units of the Earth's Crust. *Geological Society American Bulletin*, Vol. 72, pp. 175-192.

Ure, A.M., Quevauviller, P., Muntau, H. & Griepink, B. (1993). Speciation of heavy metals in soils and sediments: an account of the improvement and harmonization of extraction techniques undertaken under the auspices of the BCR of the Commission of the European Communities. *International Journal of Environmental Analytical Chemistry*, Vol. 51, pp. 135–151.

Valette-Silver, N.J. (1993). The use of sediment cores to reconstruct historical trends in contamination of estuarine and coastal sediments. *Estuaries*, Vol. 16, pp. 577-588.

Varekamp, J.C. (1991). Trace element geochemistry and pollution history of mud flats and marsh sediments from the Connecticut River estuary. *Journal of Coastal Research*, Vol. SI11, pp. 105-124.

Vilas, F., Nombela, M.A., García- Gil, E., García Gil, S., Alejo, I., Rubio, B. & Pazos, O. (1995). *Cartografía de sedimentos submarinos, Ría de Vigo*. E: 1:50000. Ed. Xunta de Galicia, 40 p.

Vilas, F., Bernabeu, A.M., Rubio, B., & Rey, D. (2010). Estuarios, rías y llanuras de marea, In: *Sedimentología. Del proceso físico a la cuenca sedimentaria*, A. Arche (Ed.), 619-673, CSIC, ISBN 978-84-00-09148-3, Madrid, Spain.

Vilas, F., Bernabeu, A.M. & Mendez, G. (2005). Sediment distribution pattern in the Rias Baixas (NW Spain): main facies and hydrodynamic dependence. *Journal of Marine Systems*, Vol. 54, pp. 261–276.

Vilas, F., Rey, D., Rubio, B., Bernabeu, A.M., Méndez, G., Durán, R. & Mohamed, K. (2008). Los fondos de la Ría de Vigo: composición, distribución y origen del sedimento, In: *Una aproximación integral al ecosistema marino de la Ría de Vigo*, A. González-Garcés, F. Vilas & X.A. Álvarez (Eds.), 17-50, Instituto de Estudios Vigueses, Vigo, Spain.

Wangersky, P.J. (1986). Biological control of trace metal residence time and speciation: A review and synthesis. *Marine Chemistry*, Vol. 18, pp. 269-297.

Wedepohl, K.H. (1971). Environmental influences on the chemical composition of shales and clays, In: *Physics and chemistry of the Earth*, L.H. Ahrens, F. Press, S.K. Runcorn & H.C. Urey, (Eds.), 307-331, Oxford, Pergamon.

Wedepohl, K.H. (1991). The composition of the upper Earth's crust and the natural cycles of selected metals: metals in natural raw materials; natural resources. In: *Metals and Their Compounds in the Natural Environment*, E. Merian (Ed.), 3-17, Weinheim, VCH.

Zoller, W.H., Gladney, E.S., Gordon, G.E. & Bors, J.J. (1974). Emissions of trace elements from coal fired power plants, In: *Trace Substances in Environmental Health*, D.D. Hemphill (Ed.), Vol. 8., 167-172, Rolla, University of Missouri, Columbia.

6

Fluorescently Labeled Phospholipids – New Class of Materials for Chemical Sensors for Environmental Monitoring

George R. Ivanov[1], Georgi Georgiev[2] and Zdravko Lalchev[2]
[1]*Department of Physics, Faculty of Hydraulic Engineering, University of Architecture, Civil Engineering and Geodesy & Advanced Technologies Ltd.,*
[2]*Department of Biochemistry, Faculty of Biology, Sofia University, Sofia, Bulgaria*

1. Introduction

Reliable environmental monitoring strongly depends on the quality of chemical and biochemical sensors. There are still some unsolved problems especially when higher selectivity is required. In this chapter we propose a new class of materials - fluorescently labeled phospholipids, which can be used as chemical and biochemical sensors. We focus our attention on the most promising compound - head labeled with nitrobenzoxadiazole (NBD) phosphatidyl ethanolamines. We were the first to study these compounds in one component layers. Three new phenomena were discovered for this material that can be used for successful sensor applications. In our research we use the Langmuir and Blodgett method for investigation of organic monolayers at the air-water interface and for thin film deposition. It can also independently be used for environmental monitoring, e.g. water purity monitoring.

2. The Langmuir and Blodgett method - Use of the Langmuir film method for measuring the quality of water in natural basins

Probably the most promising method for the creation of supramolecular architectures in a well controlled manner is the method of Langmuir and Blodgett. This method is schematically described on Fig. 1. A trough, usually manufactured from well cleanable and inert material Teflon® (polytetrafluorethylene) is filled with ultra pure water. The organic substance to be investigated and deposited is spread from a solution. Molecules of the substance should be with the proper hydrophilic–hydrophobic balance so they remain at the air-water interface and do not penetrate the water. These molecules consist of a hydrophilic head group, which is attracted to the water and a hydrophobic tail (most often - hydrocarbon groups) which is repelled by the water. Some time is allowed for the solvent to evaporate until something like a 2D gas of the investigated molecules remains at the air-water interface. This is called Langmuir film. After this a compression of the organic monolayer with a barrier is started.

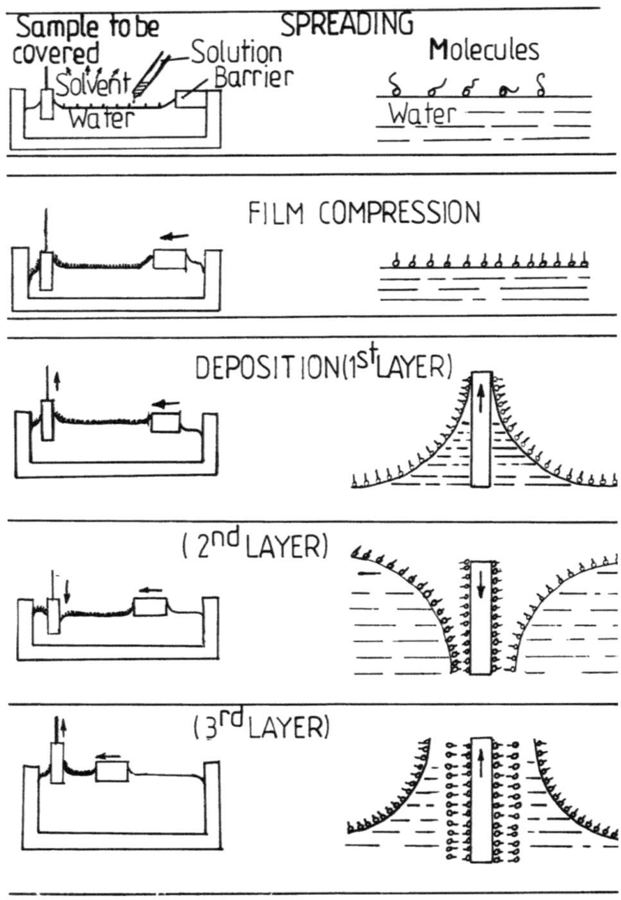

Fig. 1. The Langmuir and Blodgett method for investigation of organic monolayers at the air-water interface (Langmuir films) and for thin film deposition.

The surface pressure is constantly measured by a surface tensiometer (not shown on Fig. 1). A surface pressure - mean area per molecule isotherm can be measured. Additionally, the trough can be integrated with another instrument, e.g. fluorescence microscope and additional data can be gathered. At any point the compression of the monolayer can be stopped, the regime of constant surface pressure maintenance can be switched on, and a deposition on a solid substrate can be started. If we have a hydrophilic substrate which attracts the molecules' heads and it is immersed in the water before the monolayer spread then on the first movement up the first monolayer is deposited. Now the substrate becomes hydrophobic because the hydrophobic molecular tails are on the surface and on a downward movement of the substrate a second layer is deposited. Again the substrate surface becomes hydrophilic and on a subsequent movement upwards a 3rd layer is deposited. And the layer by layer deposition can continue. This deposition method has the following advantages compared to alternative methods of thin film deposition like spin coating, vacuum evaporation and self-assembling:

- This is a discreet method of deposition, a complete layer after complete layer are deposited. This gives the possibility for a very precise control of the film thickness. Phospholipid molecules are with height of around 3 nm but also the tail chain length can be varied so a thickness control to 0,1 nm accuracy can be achieved.
- The molecules are well oriented. This is very important for some applications, e.g. non-linear optics.
- The molecules are prearranged on the water surface before the deposition process. Thus the surface density of defects is much smaller.
- This is the most suitable method for molecular architecture. Different layers can be from different molecules. Inside the layer mixture of different molecules can be used. There is a possibility for interface reactions (e.g. CdSe nanoparticles incorporated in the lipid matrix can be prepared in this way). Also absorption from the water subphase of e.g. proteins is possible.

The development of rapid, economical and sensitive techniques for characterization of the purity of natural and drinking water represents leading ecological problem. The surface properties of natural waters (sampled from rivers, lakes and gulfs) are already successfully used for evaluation of the ecology standard and purity of water basins. Recently Pogorzelski et al. (e.g. Pogorzelski and Kogut, 2003 and references there in) proposed a Langmuir monolayer based technique which by measuring the surface pressure-area isotherm of the samples collected from a range of natural water basins, yields the so called "structural signatures" of water, which adequately predicted the quality and the purity of the basin. Major advantage of the Langmuir monolayer technique is that it combines ease of use, high sensitivity and possibility for rapid application with much lower price in comparison with the most commonly used chromatography techniques.

The "structural signatures" of samples of natural water result from the generalized scaling procedures applied to the surface pressure-area isotherms of the natural films. They appear to reflect in a quantitative and sensitive way the film composition, film solubility and the miscibility of its components, the kinetic mobility of surfactant molecules, and the compound's surface concentration. It is suggested that certain classes of film-forming components or "end-members" may dominate the static and dynamic surface properties. Variation in the surface rheological parameters of source-specific surfactants is postulated to reflect organic matter dynamics in natural waters. The reported results demonstrate that natural films are complex mixtures of biopolymeric molecules covering a wide range of solubilities, surface activities and molecular masses with a complex interfacial architecture.

The natural water's (sea, lakes, rivers) surface microlayer plays an important role in air-water interactions. A certain fraction of dissolved organic matter in the water basins has surface-active properties and makes up a very reactive part of the organic matter (Druffel and Bauer, 2000). According to their surface-active properties, these substances accumulate at water interfaces thereby influencing gas, mass, momentum and energy transfer between the so modified interfaces. The intensity of the film-effect depends strongly on film surface concentration, composition, and viscoelastic properties of the surface microlayer films. Processes taking place in the water body bulk (biological event, organic matter transformation or degradation, anthropogenic effluents, etc.) are sources of surface-active substances. Surfactants are concentrated at the air–water interface by numerous physical processes including diffusion, turbulent mixing, bubble and particle transport, and convergent circulations driven by wind, tidal forces, and internal waves.

The composition of the natural water's surface films is largely undefined, although significant enrichments of many specific classes of compounds in the surface microlayer have been demonstrated (for review, see Hunter and Liss, 1981). Natural sea/river/lake films mostly resemble layers composed of proteins, polysaccharides, humic-type materials and long chain alkanoic acid esters (Van Vleet and Williams, 1983). The generally accepted view is that the ubiquitous background of degraded biopolymeric and heterogeopolymeric material in the bulk waters has the potential to generate measurable surface films even in oligotrophic waters. Specific inputs of fresh bioexudates and biopolymeric material from local events are superimposed on this background signal.

The emphasis in the published studies (Pogorzelski, 2001; Pogorzelski and Kogut, 2001 a, b; 2003 a, b) has been on the multicomponent character of natural surfactant films and the consequent complexities involved in any attempt to predict the interfacial viscoelastic properties (playing a crucial role in modeling of physical systems with surface film-mediated interfaces) due to the diverse chemical composition of such films.

A complete compositional or structural description of naturally occurring surfactants is not currently feasible.

Instead of analyzing the chemical composition, it should be possible to scale microlayer film surface pressure-area isotherms in terms of the structural parameters, reflecting the natural film morphology, and resulting from the generalized physical formalisms adopted to multicomponent surfactant films. Particularly efficient approach to scale the surface pressure (π) - area (A) isotherms of the microlayer films adsorbed on the surface of natural waters proved to be the fitting of the isotherms by the virial equation of state as proposed by Barger and Means (1985):

$$\pi A = C_0 + C_1\pi + C_2\pi^2 \tag{1}$$

where C_0, C_1, C_2 are virial coefficients, and A is the film area (in cm^2).

As demonstrated by Pogorzelski, 2001; Pogorzelski and Kogut, 2001 a, b, 2003 a, b C_1 can be interpreted as the limiting specific area occupied by the molecules in the film, and C_o can be assumed equal to XnkT in the limiting case when π approaches zero:

$$C_0 = XnkT \tag{2}$$

where the parameter X is related to the interaction forces between molecules in the monolayer, n is the number of molecules in the unknown film, k is the Boltzmann constant, T is the temperature in degrees Kelvin.

The limiting specific molecular area A_{lim} (in nm^2) can be expressed as (Frew and Nelson, 1992):

$$A_{lim} = C_1 n^{-1} \times 10^{14} \tag{3}$$

Since the area covered with a film of a pure substance at a constant value of k is directly proportional to the mass m on the surface, it is possible to extend this computation to all the natural films (Barger and Means, 1985).

Similarly, fitting procedures can be applied to quantitatively analyze the hysteresis of natural water's surface films when subjected to cyclic area compression/expansion, and also to describe the sample's surface pressure–temperature isochores.

Thus it is possible to avoid the expensive, time consuming and cumbersome analysis of the chemical composition of natural waters, and instead to characterize the sample's quality by the introduction of sensitive and much easier to obtain physicochemical "structural signatures" of the natural microlayer films. The structural state of natural water films, which can be incorporated with such source-specific markers of both biogenic and anthropogenic origin, can be assessed through the quantification of the parameters variability. They can be useful for tracking organic matter dynamics, as already established factors like the carbon to nitrogen C/N ratio (Bock and Frew, 1993), used in microlayer film studies, for instance. The main expectation of such studies is that variation in the surface rheological parameters of natural biosurfactant films manifested at the air–water interface could be followed to trace and map surface-active source-specific compounds spatial-seasonal-temporal evolutions.

Compared to the evaluation of the ecological quality of natural waters, the characterization of the purity of drinking water poses higher challenges as it requires precise identification of even trace amount of detrimental ingredients. Some compounds, like the ions of heavy metals or membrane-active molecules, can have profound detrimental effect on the consumers' health even in very low doses that can not be detected by direct measurement of the sample's surface pressure. More precision quantitative measuring techniques are needed. For the purpose of such demanding measurements we further advanced the monolayer technique by introducing the use of fluorescently labeled LB solid supported phospholipid films. This is because fluorescence in some molecules is highly sensitive to even most delicate environmental changes (like slight changes in ionic strength, presence of quenchers in trace concentrations, etc.) LB films from these materials have high potential to be used as sensitive, selective and fast chemical sensors. In this new class of compounds for sensor applications - fluorescently labeled lipids, fluorescence intensity and lifetime are strongly influenced by minimal amounts of tested substances. In the following chapter we look in greater detail to the most promising fluorescence label in this class of compounds – the NitroBenzoxaDiazole (NBD) label. Then, results from our research of NBD labeled phospholipids at the air-water interface, as LB film on solid support, and molecular modeling are presented and discussed in view of sensor applications.

3. Previous research of NBD fluorescently labeled lipids

Synthesis, where to the polar head (to the amino group) of egg phosphatidylethanolamine (PE) covalently is bound the NBD chromophore was first described by Monti et al. (1978). Due to the use of egg phosphatidylethanolamine (PE) tail length varies. Solution of NBD-PE in ethanol shows absorption maxima at about 330 nm and 460 nm, and the fluorescence maximum is at 525 nm. Fluorescent intensity in ethanol is proportional to the concentration in the range of 1 ng/ml to about 3 μg/ml. This article studied the dependence of the intensity of absorption and fluorescence of NBD-PE to the change in dielectric constant of the solvent used. The observed strong sensitivity of the spectral characteristics of NBD-PE to the polarity of its surrounding makes this molecule an excellent indicator of conformational changes in the membrane. This article notes that small amounts of non-ionic detergent can lead to increase in fluorescence intensity and peak position change. Without problems is the incorporation of NBD phospholipid molecules in liposomes and biological membranes. For the NBD chromophores the angle between absorption and emission dipole is about 25 ° (Thompson et al., 1984) and therefore the real environment of the chromophores may be different for absorption and emission. Overview of the spectral characteristics of NBD was made by Suzuki and Hiratsuka (1988).

There are a large number of papers in which NBD labeled lipids are used especially as a small percentage additive in the biomembrane studies. Here we will review only the work related to the chemical sensor applications of these molecules. The presence of large paramagnetic metal ions can be monitored by the fluorescence quenching of the NBD chromophore. Morris et al., 1985 used cobalt ions to quench the fluorescence of NBD-PE incorporated in phospholipid liposomes. Large paramagnetic ions such as Co^{2+} efficiently quench the fluorescence. The mechanism that is suggested is of lateral diffusion of Co-lipid complex followed by collisional quenching with NBD-PE. The addition of the chelator EDTA restores the initial fluorescence to 90%. EDTA quenches itself about 10% of the fluorescence. Fluorescence is quenched in the outer layer of the liposomes within milliseconds after the addition of cobalt ions, then, if possible, it penetrates the inner layer. For small monolayer liposomes the process is 10-20 times slower, but in all cases completed in the first few seconds. This technique is used also for measuring the surface potential of the membrane. Another paramagnetic ion copper Cu^{2+} is also used for NBD fluorescence quenching (Rajarathnam et al., 1989). Chattopadhyay and London (1987) proposed a method for measuring the position of NBD chromophores in the biomembrane by quenching its fluorescence by spin-labeled in a different position phospholipids. A comparison of the fluorescence intensity is made when two located in different depths quenchers are used. Results show that the greatest distance from the center of the bilayer is for NBD chromophores in the molecules of the Dipalmitoyl-NBD-PE – 1,42 nm. This means that due to its strong hydrophilicity the NBD chromophore is folded to the hydrocarbon tails and is positioned on the border tail - head, which is 1,5 nm from the center of the bilayer. For 6-NBD-PC this distance is 1,22 nm, for 12-NBD-PC, this distance is 1,26 nm, i.e. the tail in which is the NBD chromophore is folded and goes to the water surface. In this paper is calculated the critical distance R_c, below which the fluorescence of NBD is effectively quenched by the spin-label – 1,2 nm. Calculations show that if fluorescence is quenched due to presence of acceptor this distance is 10% larger.

Another important characteristic of the NBD chromophore that can be used in sensor applications is the dependence of its fluorescence lifetime on the polarity of the surrounding media. In general, reducing the polarity of the environment increases the lifetime. Lifetime of dilauroyl and dimiristoyl-NBD-PE in liposomes of egg lecithin is 6-8 ns (Arvinte et al., 1986). Detailed analysis of the fluorescence lifetime characteristics of NBD-aminohexane acid (NBD-$NH(CH_2)_5C0_2H$) at low concentrations in solvents of different polarity and donor hydrogen connection strengths was conducted by Lin and Struve, 1991. This substance has aminoalkane side chain similar to chains in which NBD chromophore is conected to phospholipids and the results are comparable. The conclusions are that the line shift of absorption and luminescence is due to the polarity of the solvent, while the drop in luminescence intensity due to non radiation transitions is much more affected by the hydrogen connection strengths. Fluorescence lifetimes in aprotic solvents is from 7,37 ns in DMSO to 10,6 ns in ethyl acetate, but are shorter in alcohols (5,65 ns in methanol). Extremely fast is the NBD luminescence in water - 0,933 ns. Low quantum yield in water is explained by anomalously short lifetime of non radiation transitions combined with radiation transitions which are with 3 times longer lifetime than those in other solvents. Oida et al., 1993 developed the so-called Fluorescence Lifetime Imaging Microscopy (flimscopy) which uses DP-NBD-PE and rhodamine labeled lipids.

Fluorescent transduction of changes in the structure of the lipid membranes shows properties necessary for biosensor applications. When connected with the substrate a single membrane associated "receptor" protein may affect a significant number of surrounding molecules via electrostatic interactions, spatial interactions, interface changes in ionic strength or pH. The result is that: 1) perturbation of the lipid layer that is caused by the interaction receptor - ligand can be qualitatively related to the degree of connectivity, and 2) have amplified the original signal after the interaction of biomolecules. Placing a "receptor" protein in the phospholipids layer, which simulate the biological membrane, and provides improved stability against denaturation of the protein, gives biosensors with improved operational life span. Mixed lipid monolayers containing small amounts of DP-NBD-PE were shown to be able to convert changes in pH due to the hydrolytic enzyme activity at the membrane interface. This conversion scheme is used to determine the acetylcholine by acetylcholinesteraze (Brennan et al., 1990) and urea by ureaze (Brennan et al., 1992, 1993). In these studies a small concentration (about 1 mol %) of DP-NBD-PE and the respective enzyme are added in the phospholipid membrane. Changes in interface pH caused by hydrolytic enzyme reaction, lead to a change in the ionization of acidic phospholipid heads. This causes a change in the forces of electrostatic repulsion between neighboring heads. Structural changes in the membrane lead to an analytical signal in the form of change of fluorescence intensity due to fluorescence selfquenching of NBD-group caused by local increase in concentration. A comparison of different fluorophores connected to the same position of a protein showed that NBD-group gives the highest sensitivity (typically 4 times better than the next fluorophore (see Brennan et al., 2000 and references therein).

From the viewpoint of sensor applications of DP-NBD-PE important is the optimization of: a) the concentration of DP-NBD-PE molecules in the membrane, and b) the composition and structure of the phospholipids membrane. This is done by the Krull's group in Toronto (Brown et al., 1994 and Shrive et al., 1995). The results are applicable to both LB film layers and liposomes. Fluorescent measurements were performed on liposomes because the fluorescence signal from LB monolayers is weak and leads to significant errors. For the optimization process a model was developed for the fluorescence selfquenching of DP-NBD-PE. It considers the probability for static quenching by the formation of emissionless traps consisting of pairs of statistical DP-NBD-PE molecules which are at critical distance R_c. The model also considers the dynamic quenching due to Förster transfer of energy from DP-NBD-PE monomers to the traps. Assumptions in this model are: 1) statistical traps are formed according to two-dimensional equation of Perrin; 2) all DP-NBD-PE molecules that do not participate in the traps are uniformly distributed throughout the monolayer; 3) there is no diffusion during the lifetime of the excited state, 4) energy can move between and among fluorophores and traps, but once traps are reached energy immediately and without emission decreases; 5) passing of energy in more than one DP-NBD-PE molecule before reaching the trap is negligible. It is estimated that the distance at which the efficiency of Förster transfer of energy becomes 50% R_0 = 2,55 nm and that R_c = 0,94 nm. The optimum concentration of DP-NBD-PE molecules is one in which the theoretical expression undergoes a maximum change, i.e. the second derivative of the expression to the change in concentration is calculated. According to theoretical calculations, the optimal concentrations were 0,027 and 0,073 DP-NBD-PE molecules per nm^2. These values were the same within the experimental error when comparing results of three different types of liposome compositions.

Optimization of composition and structure of membrane phospholipids showed the need for structural heterogeneity in the membrane at microscopic and not at molecular level in order to produce significant changes in fluorescence intensity. In membranes without heterogeneity the signal change is only 5-6%. Heterogeneity is achieved by the mixing of dipalmitoyl phosphatidyl choline with dipalmitoyl phosphatidic acid at a ratio of 7:3. At surface pressure of 30 mN/m, which is considered the liposome pressure, this mixture gives domain structure as observed in Langmuir films by fluorescence microscopy. The resulting changes in the average fluorescence intensity on pH change in this case reaches 60%. The mechanism of response of the membrane is shown to depend on the surface potential (Nikolelis et al., 1992) and is the result of changes in the ionic double layer and the rearrangement of the lipid heads and tails. This indicates that the mechanism of response in these biosensors is much more complicated than changing the distance between the heads. Moreover, the choice of phospholipid for these biosensors must be based on constraints coming from the ionic strength and pH, imposed on the activity of immobilized, chemically selective protein as enzyme activity is highly dependent on pH (Brennan et al., 1994).

4. Investigations of monolayers at the air-water interface

On Fig. 2 are shown the isotherms of head labeled Dipalmitoyl-NBD-PE (DP-NBD-PE, chemical formula is in Fig. 14) at three different temperatures and in the presence of cobalt ions in the water at 20° C. Along with these measurements the monolayer was studied with fluorescence microscopy. The results for 20° C are published and discussed in detail elsewhere (Ivanov, G.R. (1992)). This was the first time that fluorescence self quenching in organic monolayers at the air-water interface (Langmuir films) was described. Here on Fig. 3 for the first time we publish the fluorescence microscopy data at 5° C. At room temperature the average area per molecule in the liquid phase is 1,4 nm^2, and in the solid phase - 0,45 nm^2. Adding $CoCl_2$ in the water increases the surface area of the molecule in the solid phase to 0,67 nm^2. The addition of $CaCl_2$ (not shown) leads to a smaller increase in the area in the solid phase - 0,56 nm^2.

The shape of the solid domains is due to an interplay of several forces: the growth kinetics which at these compression speeds is negligible; the edge energy at liquid phase – solid phase interface which is minimal for circular domains; and the electrostatic repulsion between the similarly oriented dipoles of the molecules which is minimized when the molecules are further apart. Due to the last force the domains repulse each other at low surface pressures and when the area of the solid domain is increased at higher pressures the domains obtain the dendridic shape which increases the distance between molecules. The fluorescence microscopy data at 5° C reveals also something that is not well observed at higher temperatures. The solid domains grow in size largely due to the attachment of smaller solid domains from the second population of solid domains.

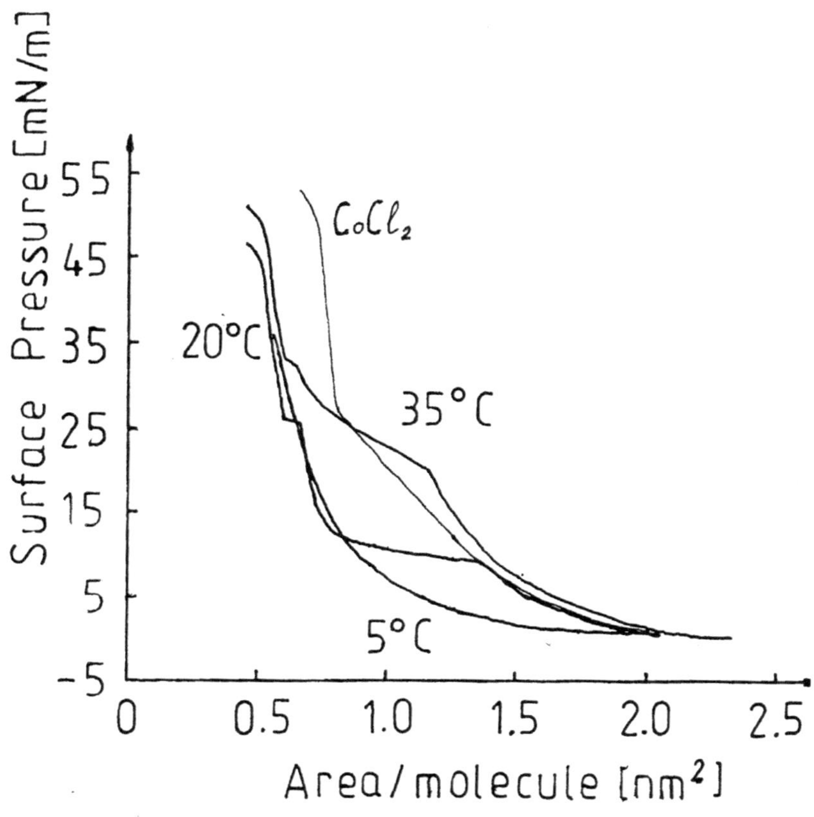

Fig. 2. Isotherms of monolayers from DP-NBD-PE at 5° C, 20° C, 35° C and at 20° C with the presence $CoCl_2$ in the water.

The presented data here is for single component monolayers composed only from the fluorescently labeled in the head phospholipid DP-NBD-PE. The fact that we are able to observe the picture of phase coexistence with an excellent contrast is due to the fluorescence self quenching of this molecule in the solid phase when the distance between the molecules becomes much smaller and this allows for non radiation transfer of energy between them. This new phenomenon can be used with great success in sensor applications. If due to interactions of the sensor with the substance to be detected some conformational changes in the DP-NBD-PE molecules arise, this will lead to a strong measurable change in the fluorescence intensity. So this provides a second mechanism for component detection apart from the already discussed influence of the fluorescence peak maximum, intensity and lifetime on the polarity of the surrounding medium.

On Fig. 4. is shown the equilibrium spreading pressure measurement of DP-NBD-PE at 20° C. In the first few seconds after placing some crystals from the material on the water surface the surface pressure increases insignificantly, then within a few seconds it increases by more than 15 mN/m. Then within a minute it reaches its equilibrium value of 19,6 mN/m. This value is quite high and indicates that at room temperature the majority of studies described in this work were conducted under equilibrium conditions. This is not quite so with much of the work conducted with LB films. For example the most widely studied arachidic acid has an ESP of 0 at room temperature indicating that the molecules are in metastable state when deposited.

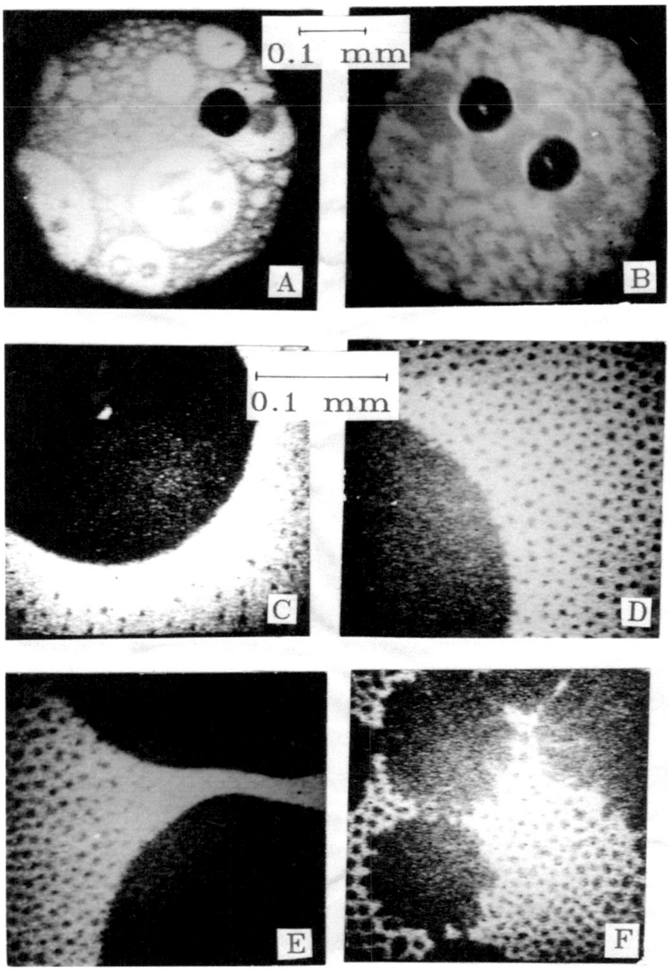

Fig. 3. Fluorescence microscopy of monolayers at the air-water interface from DP-NBD-PE at 5° C at different surface pressures Π: (A) Coexistence of liquid and solid phase at zero pressure; (B) the same at Π = 0,2 mN/m; (C) the occurrence of a second population of solid phase (the small black dots), which is repelled from the large solid domains Π = 6 mN/m; (D) the small solid phase domains overcome the repulsion of the big domains and begin to attach to them at Π = 9 mN/m; (E) large domains close the distance between them at Π = 12 mN/m; (F) domains of the solid phase obtain the dendridic shape and begin to merge with each other at Π = 15 mN/m.

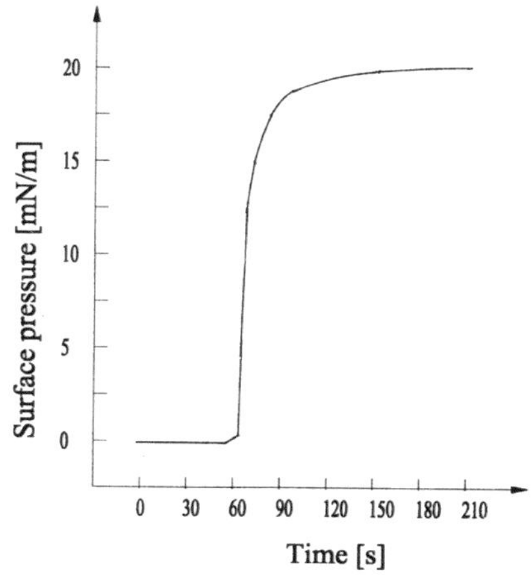

Fig. 4. Measurement of the equilibrium spreading pressure (ESP) of DP-NBD-PE at 20° C.

5. Investigations of deposited on solid support thin films using the LB method

For the possible sensor applications of DP-NBD-PE molecule it is important to obtain quality deposition on solid support. Deposition of multilayer structures of phospholipids by the LB method is complicated. Usually when immersed for the second layer deposition the first one is thrown off back into the water probably due to the presence of residual water between the layer and the substrate. A similar phenomenon was observed in the case of DP-NBD-PE. A typical way to overcome this problem is to use extremely low deposition speed. So no water is entrapped and relatively high quality multilayer structures are obtained. The problem is that very few commercial instruments have such low deposition speeds and these speeds are not suitable for industrial applications. Therefore we proposed a new method of obtaining multilayer structures. In it, after each immersion the bilayer was blown with heated air for several minutes at 55°C, which is below the melting temperature of the monolayer. The deposition results are compared in Table. 1. The quality of the film, or more precisely the amount of transferred substance, is judged by two criteria. On one hand this is the transfer ratio (Tr), which ideally is 1. But its determination has large errors due to difficulties in maintaining a constant surface pressure, causing the barrier to move back and forth without much correlation with the deposited layer. Far more accurate method is the measurement of the optical absorption of the film. In it the area of the line of maximum absorption of DP-NBD-PE at 465 nm is integrated.

Usually, in order to improve the LB film quality metal ions are added in the water subphase. With fatty acids good results are obtained when divalent ions of heavy metals such as Cd^{2+} are added but in our case this did not lead to good results. For phospholipids it is usually recommended the use of univalent metal ions. They bind to the negatively charged phosphate head of the phospholipid and neutralize the electrostatic repulsion between neighboring layers. With our molecules best results were obtained with the use of NaCl.

Table. 1 shows that the use of thermal treatment of the film increases the transfer rate which means that more substance is deposited. Peeling off the film on the down substrate movement is greatly reduced, although almost always the transfer ratio on the down movement is less than the coefficient in the upward movement of the substrate. The results of optical absorption also confirm that more substance is deposited when heat treatment is used.

Deposition conditions			Transfer ratios (Tr) for the corresponding layer												Integral Absorption
No	Π (mN/m)	Subphase	Tr_1	Tr_2	Tr_3	Tr_4	Tr_5	Tr_6	Tr_7	Tr_8	Tr_9	Tr_{10}	Tr_{11}	Tr average	(a. u.)
G1	20	D.W./H.	3,2	0,4	2,1	-0,3	1,9	-0,5	1,7	-0,5	1,5	-0,6	1,5	1,40	0,45
G2	31	D.W./H.	2,7	0,8	1,3	0,7	1,3	1,0	1,1	1,0	0,6	1,1	0,3	1,21	3,16
G3	40	D.W./H.	2,1	0,1	0,7	0,4	1,1	0,7	1,0	0,6	0,7	0,7	0,7	0,70	2,56
G4	43	D.W./H.	1,1	0,5	0,5	0,3	0,6	0,4	0,7	0,6	0,7	0,4	0,7	0,80	2,02
G5	47	D.W.	1,3	1,4	1,1	0,9	1,1	0,9	1,9	0,9	1,3	0,9	0,9	1,67	-
G6	31	$CdCl_2$/H.	3,2	-4,8	4,8	-11,8	11,8	-11,0	11,7	-13,2	10,6	-10,0	9,2	0,35	0,74
G7	31	KCl/H.	4,0	1,9	2,1	1,1	2,3	1,4	3,1	1,3	2,1	1,0	2,3	1,80	0,82
G8	35	NaCl	2,3	-5,7	5,8	-5,6	5,4	-4,4	5,5	-3,7	4,0	-3,7	4,1	0,36	0,76
G9	35	NaCl/H.	2,7	0,4	3,4	0,4	3,1	0,6	2,5	0,5	2,7	1,3	3,4	2,29	2,59

Table 1. Transfer ratios (Tr) and integral absorption of LB films from DP-NBD-PE. H means heat treated, DW means distilled water.

To examine the effect of thermal heat treatment on the morphology of the resulting LB films comparative optical microscopy studies of samples G8 and G9 were performed. They are deposited at the optimum surface pressure of 35 mN/m and in the presence of NaCl in the water subphase. The only difference is that in G9 each bilayer was heat treated. Results from the dark field and phase-contrast microscopies clearly showed that considerably more substance is deposited when heat treatment is used and the density of defects is significantly lower. However, a significant number of defects and distinct domain structures with dimensions of tens microns can be seen.

Another way of assessing the quality of the deposition is to measure the mass of the deposited substance. This is done when the LB film is deposited directly on a quartz crystal resonator. We used a resonator operating at 10 MHz frequency. It should be noted that sensitivity depends on the square of the frequency of the resonator. This is one of the most promising methods for creating gas and biochemical sensors which directly measures the mass of the substance to be detected. The accuracy of the measured frequency shift is below 0,1 Hz and the sensitivity of the method can be seen. It was possible to observe the water evaporation from the layer. In the middle of the resonator is evaporated a gold heater which can be used in chemical sensor applications for desorption of the absorbed studied substance. On the same resonator were sequentially deposited 21 layers, then frequency was measured, then 12 more layers were deposited and the frequency was measured, and finally 6 more layers were deposited and frequency measured (Fig. 5). In the case of an ideal deposition the mass (evaluated from the frequency change) should lie on a straight line. In our case this occurs with a deviation of around 10% indicating a high quality deposition. During this deposition every bilayer was heat treated.

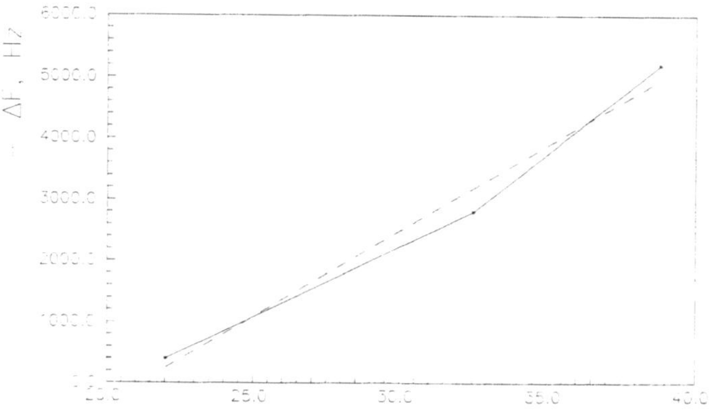

Fig. 5. Increase of the mass of 21, 33 and 39 LB monolayers from DP-NBD-PE measured with quartz resonator.

To get an idea of the effect of heat treatment at molecular level polarization Fourier transformed infrared spectroscopy in attenuated total reflection mode of multilayer LB films deposited at 35 mN/m was conducted. The results are shown in Fig. 6. The bottom curve shows the spectrum of film obtained at very low deposition speed. The middle curve shows the same film heated for several minutes at 55 ° C. The upper curve shows the spectrum of

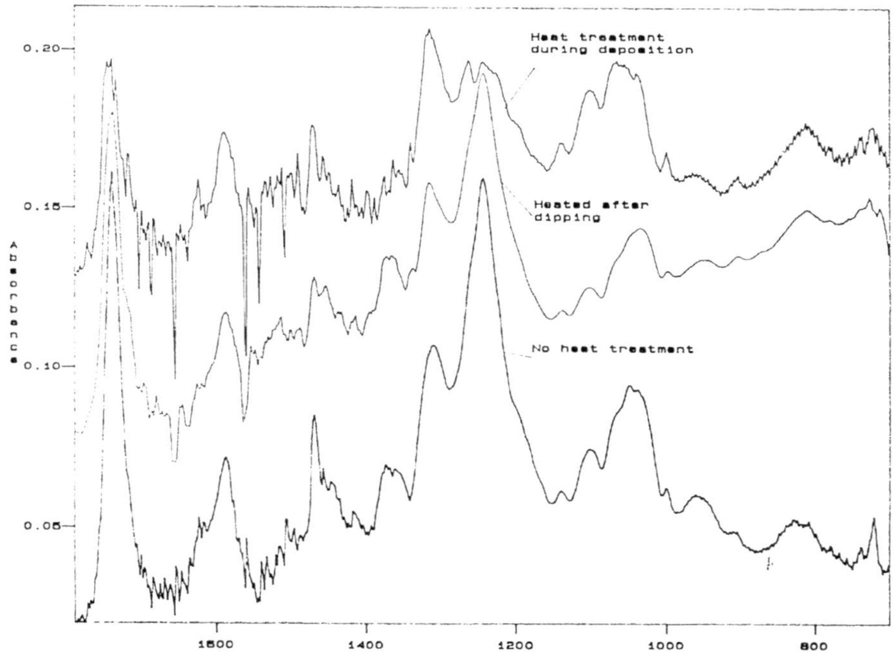

Fig. 6. The influence of heat treatment on multilayer LB films from DP-NBD-PE on the infrared spectra.

film obtained by high speed deposition and heat treatment of each bilayer during the deposition. The most important change is the significant broadening of the absorption lines at 1244 cm^{-1} (which is a mixture of lines Y_w CH_2 and Y_w P=0) and especially at 1738 cm^{-1} (which corresponds to the Ys C=0) when heat treatment is performed. Particularly noticeable is this broadening when each bilayer is heated. The broadening of these lines indicates greater spread in the orientation of the corresponding parts of the DP-NBD-PE molecule. This is an expected result when heat treatment is performed.

Additional information about the molecular arrangement and orientation of the chromophore head may be obtained from polarized absorption spectroscopy in the visible region. Simple NBD-derivatives have three main lines of absorption in the visible and near UV region - at around 420 nm, at 306-360 nm, and 225 nm (Lancet and Pecht, 1977). The first line corresponds to the line of 460 nm for NBD-labeled lipids and is due to intramolecular charge transfer (Paprica et al., 1993), which is accompanied by a large (~ 4 Debye) change in dipole moment (Mukherjee et al., 1994). The line absorption at 306-360 nm (for NBD-lipids ~ 335 nm) corresponds to a transition $\pi^* \leftarrow \pi$. Absorption spectra of NBD labeled lipids are shown in Fig. 7. The chloroform solution of a tail NBD labeled dipalmitoyl phosphatidylcholine has a maximum absorption at 475 nm. DP-NBD-PE in chloroform solution has a maximum at 457 nm. When deposited as LB film it has absorption maximum at 460 nm and a pronounced shoulder at 492 nm. This shoulder is probably due to J-aggregation of molecules in which the optical dipoles are arranged like a brick wall (Czikklely et al., 1970 a, b). J-aggregates are characterized by red shift of the absorption spectrum by about 30 nm and therefore this is the most likely interpretation. The presence of a large percentage of J-aggregates in the condensed phase in which deposition was carried out, may explain the fluorescence quenching in the solid phase.

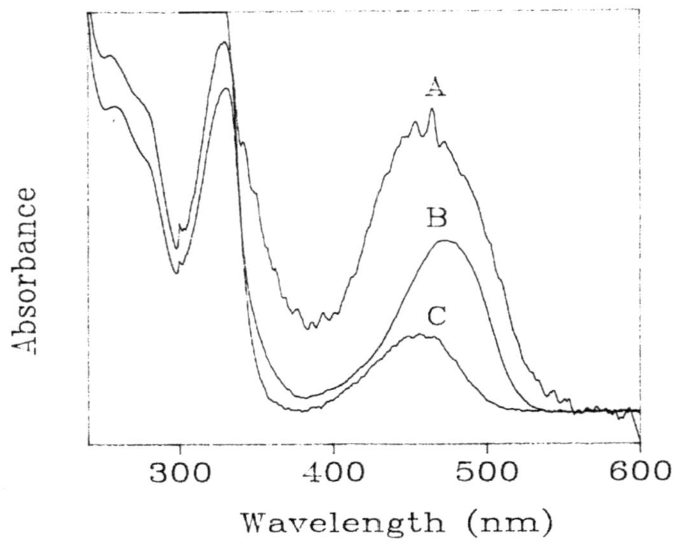

Fig. 7. UV-VIS absorption spectra of: A – LB film from DP-NBD-PE deposited at 35 mN/m and NaCl in water; B – chlorophorm solution of dipalmytoyl phosphatidyl choline labeled in the tail with NBD; C – chlorophorm solution of DP-NBD-PE.

On Fig. 8 and Fig. 9 are shown the polarization absorption spectra of two samples of LB films from DP-NBD-PE with identical thickness. Both normal incidence of the beam and incidence at 45° angle is used (index 45) to the substrate. The beam polarization is either parallel to the direction of withdrawal of the substrate (index p) or perpendicular to it (index s). Results for the area integral under the curves are summarized in Table. 2.

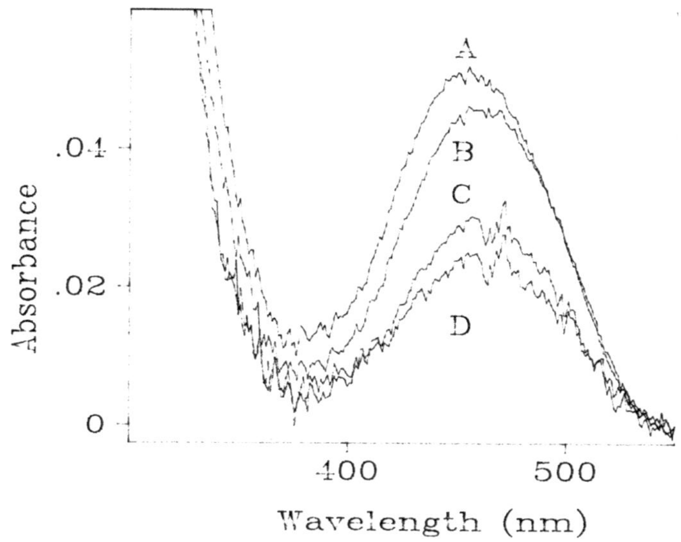

Fig. 8. Polarization spectroscopy of sample G5 - 11 LB layer structure of DP-NBD-PE deposited at Π=47 mN/m and fast withdrawal. A - p45; B - s; C - p; D - s45.

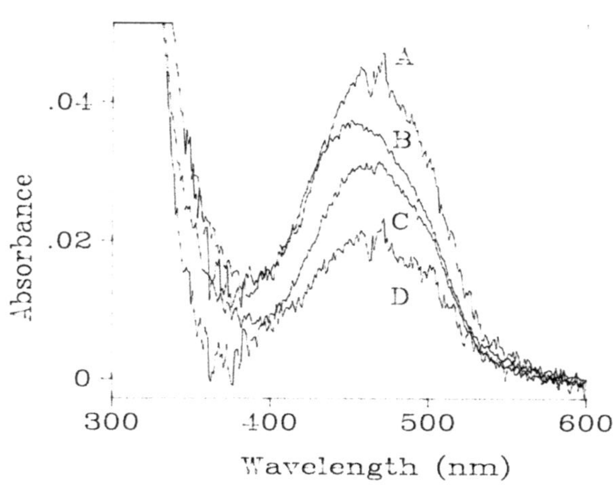

Fig. 9. Polarization spectroscopy of sample G9 - 11 LB layer structure of DP-NBD-PE deposited at Π = 35 mN/m with heat treatment and the presence of NaCl in the water. A - s45; B - p45; C - s; D - p.

LB film	p	s	p45	s45
G5	10,38	11,39	4,06	2,17
G9	1,93	2,27	2,55	3,35

Table 2. Integral areas of the 460 nm line of the absorption spectra at different polarization of the incident light for 2 different 11 layer LB films from DP-NBD-PE. G5 and G9 are the same films from Table 1. G9 was heat treated during the deposition.

From the polarization data it is seen that in both cases the optical dipoles of the molecules orient themselves with a slight advantage in the direction perpendicular to the direction of withdrawal (**s** component is larger than **p** component). From the data in table. 2 order parameter for the molecules $<\cos^2 \theta>$ can be calculated. The value for G5 film is 0.48. If the arrangement is perfect the value should be 1. We can see the weak ordering of molecules or more precisely of their chromophore heads.
Fluorescence spectrum of LB multilayer structure of DP-NBD-PE deposited on a glass substrate in the solid phase was compared with the spectrum in chloroform solution in Fig. 10. Excitation was at the maximum absorption at 465 nm. About five-fold decrease in the fluorescence in the solid phase can be seen. When deposition is carried out in the liquid phase below surface pressure of 8 mN/m fluorescence is similar to that in a solution. In solid phase deposition the fluorescence quenching is almost complete. This is not due to reduced absorption, as absorption is increased by 10% in the solid phase deposited film. Interestingly, the addition of cobalt ions in the water subphase leads to almost complete recovery of fluorescence to its level in the solution (line not shown). This can be used in chemical sensors for heavy metal detection.

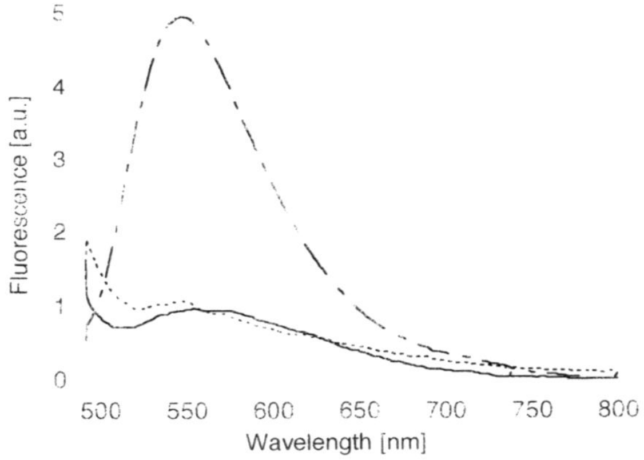

Fig. 10. Fluorescence spectroscopy of DP-NBD-PE of an LB film deposited in a solid phase (lower 2 curves) and in a solution. The spectra of an LB film deposited in a liquid phase is similar to the spectrum in solution.

An important question in view of sensor applications are the mechanisms of fluorescence selfquenching in NBD-labeled lipids. There is a similar study for octadecyl-rhodamine molecule (MacDonald, 1990), commonly used in Resonance Energy Transfer (RET) studies with NBD-molecules. The most obvious possibility for a mechanism to quench the fluorescence is the collision between an excited molecule and a quencher molecule. For this process is important the local concentration of fluorescently labeled molecules in the liquid phase. Calculations show that for collisions to occur with molecules with diffusion rate of 10 $\mu m^2/s$ and fluorescence lifetime of 4 ns then the distance between them must be less than 0.3 nm. And such small distance is impossible for molecules with 2 tails in a liquid phase, even assuming an increase in local concentration. In monolayers that undergo phase transition liquid - solid state diffusion rate decreases more than three orders of magnitude: from 50 $\mu m^2/s$ to 0,03 $\mu m^2/s$ (Peters and Beck, 1983). Fluorescence lifetime of dilauroil and dimiristoil NBD-PE in liposomes of egg lecithin is 6-8 ns (Arvinte et al., 1986). Thus, in the liquid phase diffusion during the excited state of the DP-NBD-PE is about 3 nm and in solid phase it is only about 0,02 nm. From the isotherm of DP-NBD-PE (Fig. 2) it can be seen that at room temperature the average area per molecule in the solid phase is 0,45 nm, and in liquid phase is 1,4 nm. Upon an assumption of cylindrical adjacent tightly packed molecules the distances between the centers of molecules in liquid phase is 1,33 nm and 0,78 nm in the solid phase. Obviously the collisional quenching mechanism of fluorescence is not applicable in solid phase. However, it is not clear why there was no significant fluorescence quenching in the liquid phase. The answer probably lies in the observation of Lin and Struve (1991) for extremely fast fluorescence lifetime of the NBD chromophore in water - 0,933 ns. Indeed, at the air-water interface the NBD-group in DP-NBD-PE molecule is positioned entirely in the water as shown by molecular conformational modeling (see Fig. 14). Another possible explanation is the dependence of the lifetime on the concentration of DP-NBD-PE (Brown et al., 1994). At all concentrations the lifetime is a two exponent function and the average ranged from 8,66 ns at 0,1% concentration; 5,39 ns at 10%; 1,32 ns at 40%; 0,97 ns at 50%. At 100% concentration (we work with films composed only from this molecule) the lifetime will probably be even smaller. The above mentioned distances between molecules in the liquid phase are several times larger than the distance which an excited molecule can diffuse for these small lifetimes. This explains why there is no fluorescence quenching in the liquid phase.

Therefore, as a possible mechanism for fluorescence quenching remains energy transfer. For octadecyl-rhodamine molecule the distance at which energy transfer to monomer or dimer of the same molecule is 50% (Förster radius) is 5,5-5,8 nm for a transfer to monomer and 2,7 nm for a transfer to dimer. This calculation is done using a formula that takes into account the spectral overlap of the excitation and emission spectra for a given molecule. For NBD-molecules the Förster radius R_0 is 2,55 ± 0,15 nm (Brown et at., 1994) . Research that shows anomalous long distance energy transfer (Draxler et al., 1989; Fromhertz and Reinbold, 1988) should also be taken into account. In LB films, they observed 20% efficiency of energy transfer over distances of 150 nm. Depending on mutual orientation of molecules, this distance may decrease to 30 nm. If the molecules, among which energy transfer takes place are different, then the lifetime of the donor molecule should increase with increasing its concentration. But in the case of octadecyl-rhodamine it decreases. Therefore, the basic mechanism of fluorescence quenching in octadecyl-rhodamine is emissionless energy transfer to the dimers of the same molecules. This lipid associated fluorescence label forms pre-bonded dimers. However this mechanism does not explain why fluorescence quenching in DP-NBD-PE molecules occurs only in the solid phase.

As a most probable reason for the fluorescence quenching for DP-NBD-PE a model was developed that takes into account the likelihood of static quenching by forming emissionless traps consisting of pairs of statistical DP-NBD-PE molecules that are at critical distance R_c for trap formation (Brown et al., 1994; Shrive et al., 1995). Calculations give a value for R_c of

0,94 nm for this molecule. This value is greater than the intermolecular distances in solid phase layers of DP-NBD-PE and less than the distances in the liquid phase at room temperature. The addition of cobalt ions leads to increased intermolecular distances of up to 0,93 nm, which may explain the recovery of fluorescence intensity of the solid phase in presence of this ion which is known as good fluorescence quencher. Thus, this model explains fluorescence quenching in the solid phase, the lack of quenching in the liquid phase and the recovery of fluorescence in the solid phase with the addition of cobalt ions in the water. Given the anomalous long-distance energy transfer, observed in LB films (Draxler et al., 1989; Fromhertz and Reinbold, 1988) it is clear that critical is the formation of traps, which is precisely what we observed.

In order to have a better understanding of the morphology of the LB films with submicrometer resolution we performed Atomic Force Microscopy (AFM) measurements. Fig. 11 shows AFM images with cross-sections along selected lines. On Fig. 11 A deposition was carried out at 7 mN/m, slightly above the transition from liquid to solid phase. The liquid - solid phase coexistence is clearly seen. Cross section through one of the solid domains reveals cylindrical structure with a height of 5,8 nm and a width of several tens of nanometers. Taking into account the height of the monolayer (3,1 nm), these structures can be interpreted as a bilayer in the solid phase. These structures are observed systematically in all our experiments. They occur above the main liquid-solid phase transition. Deposition in the liquid phase (Fig. 12) showed that there are no structures outside of the normal silicon wafer bumps of less than 0,5 nm in the monolayer.

Cylindrical structures with bilayer height are observed also when the depositions are carried under the equilibrium spreading pressure of 19,6 mN/m (see Fig. 4). If the deposition at higher pressures is carried out these cylinders grow in height initially up to 13 nm for deposition at 33 mN/m (Fig. 11 B) and grow up to 35 nm, and in some cases to a hundred nanometers at 43 mN/m (Fig. 11 C). However, if the monolayer is allowed to relax under normal laboratory conditions for some time (50 days in this case), those cylinders again become with bilayer height (Fig. 11 D).

These three-dimensional cylinders can not be obtained due to the deposition process and/or interactions with the substrate because their height depends on the deposition pressure. Layering in the vertical direction of the DP-NBD-PE molecule can be observed in these high structures, suggesting that the cylinders are made of DP-NBD-PE molecules rather than impurities. The question remains why it is energetically more favorable for part of the molecules to accumulate on one another and not to attach to the adjacent solid phase. Possible answer is that this can be due to kinetic effects if we are compressing the layer or depositing at higher speeds. Against this explanation is the lack of such structures in the liquid phase deposition under the same conditions (Fig. 12). Furthermore, our previous data from fluorescent microscopy at the air-water interface does not show the presence of kinetic effects at these speeds. Against this explanation is the fact that high structures relax to bilayer cylinders with time (Fig. 11 D). Data from Stark spectroscopy measurements show that DP-NBD-PE molecules tend to form centrosymmetrical non-polar structures. Thus for the second layer in the bilayer structures the molecules most probably flip over and we have a tail - tail contact. This is the first time that such 3D structures are observed when deposition is carried below the ESP. These structures are stable over time at least for several months. Their presence is very important for sensor applications because they ensure simultaneously high contact area and low film thickness. Thus high sensitivity at fast reaction times can be achieved (no slow diffusion needed). These self assembled 3D structures are part of our efforts in the hybrid assembly approach which combines the self-assembly technology with high performance robotic tools such as precise manipulators with submicron resolution and mechatronic handling (Kostadinov, 2010; Dantchev and Kostadinov, 2006).

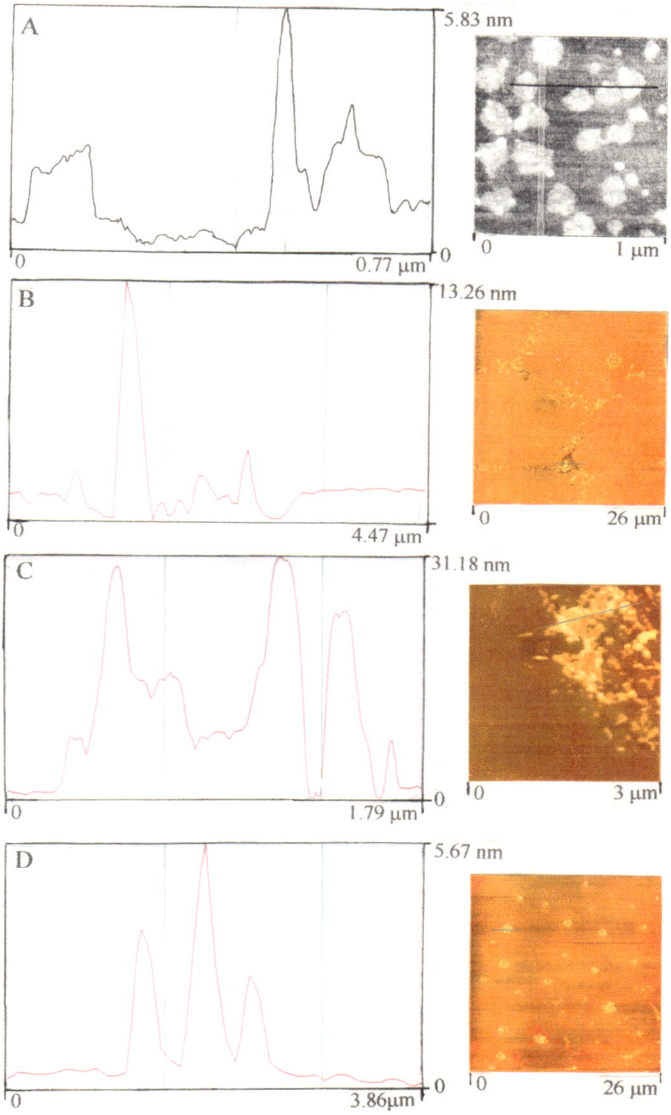

Fig. 11. AFM pictures and crosssections of LB monolayers from DP-NBD-PE deposited from pure water at 20^0 C and surface pressures of: (A) 7 mN/m; (B) 33 mN/m; (C) 43 mN/m; (D) 33 mN/m 50 days after the deposition.

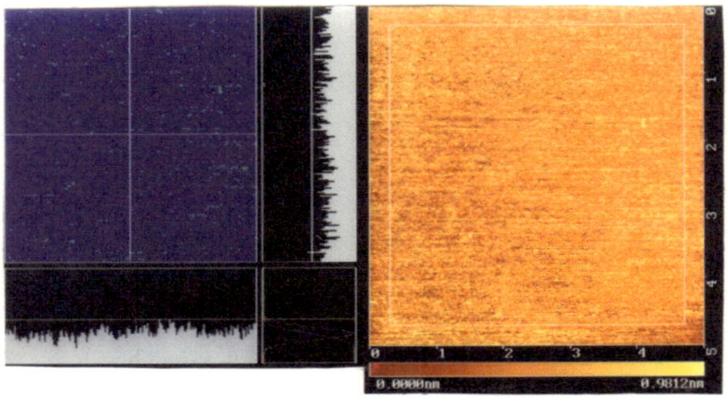

Fig. 12. AFM pictures and crosssection (left at a level of 1,04 nm) of LB monolayers from DP-NBD-PE deposited from pure water at 20^0 C and surface pressures of 3,7 mN/m. Scan size is 5 μm.

6. Molecular modelling of the DP-NBD-PE molecule at simulated air-water interface

In order to get a better understanding of the 3 newly discovered phenomena in LB films from DP-NBD-PE we have performed molecular conformational analysis at simulated air-water interface. The two most probable conformations from this analysis are shown on Fig. 14 and their characteristics are summarized in Table. 3.

Conformation	Height [nm]	phi - pho distance Δ [nm]	Area per molecule – single molecule [nm^2]	Area per molecule – in a monolayer [nm^2]
A - liquid phase	2,31	0,938	1,18	0,81
B - solid phase	3,25	1,396	-	0,69

Table 3. Characteristics of the two most probable conformations for the DP-NBD-PE molecule at the air-water interface. The conformation in the solid phase is obtained only when interactions with surrounding molecules are taken into account.

The solid phase conformation A data correspond very well to the experimental data from the previous paragraphs. The average area per molecule is 0.66 nm^2 if the isotherm from Fig. 2 is extrapolated to zero surface pressure, which almost coincides with the measured value of 0,69 nm^2. The height of the molecule measured with different methods (including small angle X-ray diffraction) is 3,1 nm, which taking into account for some interdigitation of tails in the LB multilayer structure, matches the obtained here value of 3,25 nm. Also the predictions from the measurements of molecular orientation from polarized FTIR data are fulfilled. Particularly impressive is that the benzene ring is indeed perpendicular to the substrate.

The liquid phase conformation in Fig. 13 A corresponds to the liquid film of DP-NBD-PE provided that the tails are even flatter and not in *all-trans* conformation. Indeed, the area per molecule in the liquid phase at zero pressure is about 2 nm, which is more than the 0,81 nm predicted by our model. Also AFM measurements showed a difference in height between the liquid and solid phase of about 1,6 nm. If the height of the molecule in the solid phase of 3,1 nm then for the height of the molecule in liquid phase remains 1,5 nm, which is less than the 2,31 nm in our model. So the assumption of greater tilting of the tails, leading to a lower height of the molecule and larger area is fully justified. Scanning surface potential microscopy measurements show that there is a big difference in the surface potential of the monolayer in liquid and in solid phase, most likely due to the different orientation of the strong dipole in the NBD-group. Indeed, the orientation of this group for the conformations on Fig. 14 show an angle of almost 90 ° between them, which may explain these results. In the liquid phase conformation A just over half of the chromophores are above the air-water interface, while the NBD group in the solid phase conformation B is deeply immersed in the water. The difference in dielectric constants of the environment in both cases leads to different fluorescence quenching by the water and may explain some phenomena observed by the fluorescence microscopy.

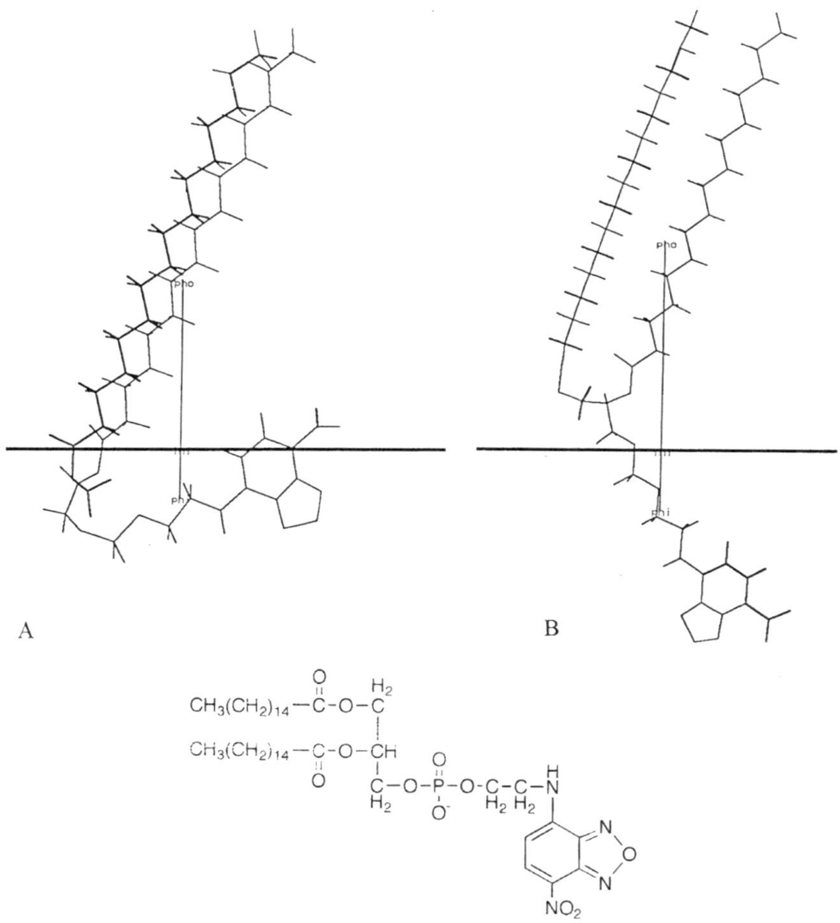

Fig. 13. The most probable conformations of DP-NBD-PE obtained from molecular conformational analysis. (A) The conformation in the liquid phase; (B) the conformation in the solid phase. Also shown are the hydrophilic (phi) and hydrophobic (pho) centers. The line that connects them is the phi-pho distance Δ in Table 3. The chemical formula of the compound is shown below.

The hydrophilic-hydrophobic balance Φ for DP-NBD-PE was calculated to be 0,55. The distance Δ between the hydrophilic and hydrophobic center in the liquid phase conformation was 0,938 nm, while in the solid phase conformation it is 1,396 nm. According to the classification of Brasseur, 1990 (vol. 1, p. 210) both conformations fall in the zone with $\Phi > 0,2$ and $\Delta > 0,43$, which is characteristic of molecules that can self assemble in organized structures. Another prediction of Brasseur is that due to the large difference in Δ of the two conformations at the molecular level they can not mix and have to form two phases. Exactly this is what we observe in our experiments and there is coexistence between liquid and solid phase and well seen phase separated domains.

7. Conclusions

We were the first to start investigating systematically films at air-water interface and on solid support from fluorescently NBD-labeled phospholipids. Previous research has shown that this is the most promising fluorophore label for sensor applications. In our investigations we use the most advanced method for preparing supramolecular architectures from organic molecules – the method of Langmuir-Blodgett film deposition and research. This is a true nanotechnology process.

Over the years we have discovered 3 new phenomena in these molecules which make them a promising candidate for chemical and biochemical sensor applications when fast response times, high sensitivity and selectivity are required. We were the first to observe fluorescence self quenching in insoluble monolayers at the air-water interface. Self quenching not only drastically decreases fluorescence intensity but also leads to a decrease in fluorescence kinetics times by an easily measurable change of over 30 %. Thus we have 2 independent channels to discriminate the effect in a sensor application. This phenomenon was understood in terms of molecular conformational change which leads to more dense molecular packing in the solid phase and radiationless energy transfer between the closely spaced molecular heads. So any change in the molecular environment which leads to this conformational change can be easily detected.

The second new phenomenon describes the influence of heavy metals on fluorescent intensity in this type of molecules. Usually large paramagnetic metal atoms are strong fluorescence quenchers. But when they are dissolved in the water subphase during the deposition process the opposite effect was observed – the fluorescence intensity was increased. This was explained by the fact that these large atoms effectively increase intermolecular distances in the head of the molecules where they attach and thus decrease the fluorescence self quenching described above. This effect can be used for heavy metal detection.

The third new phenomenon describes the possibility to deposit monolayers at some special conditions in which there is not only coexistence of solid and liquid phase but higher, bilayer or tens of nanometer high cylinders are deposited. This structure was very stable at least within several months period. It allows a much greater contact surface between the fluorescence molecules and the substances to be detected. Thus, high sensitivity sensors can be obtained without increasing their thickness. When the thickness is small so are the diffusion lengths which limit the sensor reaction time. Thus, very fast sensors with high sensitivity can be obtained. The possibility to mix selectively reacting proteins in this flexible phospholipid matrix can provide an unmatched selectivity. These properties are very important for environmental monitoring.

8. Acknowledgments

This work is supported by the following contracts with the Bulgarian National Science Fund: VU-F1/2005, DO 0171/2008, DO 02-280/2008, DO 02-107/2008 and DO 02-167/2008.

9. References

Arvinte, T., A. Cudd, and K. Hildenbrand. (1986). *Biochim. Biophys. Acta*, Vol. 860, p. 215

Barger, W.R., Means, J.C., (1985) Clues to the structure of marine organic material from the study of physical properties of surface films. In: Sigleo, A.C., Hattori, A. (Eds.), *Marine and Estuarine Chemistry*. Lewis Publishers, Chelsea, pp. 47–67

Brasseur, R. ed. (1990) *Molecular Description of Biological Membranes by Computer Aided Conformational Analysis*, CRC Press, Boca Raton

Brennan, J.D. and UJ. Krull. (1992). *Chemtech*,Vol. 22, p. 227

Brennan, J.D., R.S. Brown, A. Delia Manna, K.M.R. Kallury, P.A. Piunno, and U.J. Krull. (1993). *Sens. Act. B*, Vol. 11, p. 109

Brennan, J.D., K.M.R. Kallury, and U.J. Krull. (1994). *Thin Solid Films*, Vol. 244, p. 898

Brennan, J.D., K.K. Flora, G.N. Bendiak, G.A. Baker, M.A. Kane, S. Pandey and F.V. Bright. (2000). *Phys. Chem. B*, Vol. 104, p. 10100

Brown, R.S., J.D. Brennan, and U.J. Krull. (1994). *J. Chem. Phys.*, Vol. 100, p. 6019

Chattopadhyay, A. and E. London. (1987). *Biochemistry*, Vol. 26, p. 39

Czikklely, V., H.D. Forsterling and H. Kuhn. (1970 a). *Chem. Phys. Lett.*, Vol. 6, p. 11

Czikklely, V., H.D. Forsterling and H. Kuhn. (1970 b). *Chem. Phys. Lett.*, Vol. 6, p. 207

Dantchev D. and K. Kostadinov (2006). On forces and interactions at small distances in micro and nano assembly process In: 4M 2006 *Second International Conference on Multi-Material Micro Manufacture*, Edited by: W. Menz, St. Dimov, B. Fillon. pp.241-245, Oxford: Elsevier

Draxler, S., M.E. Lippitsch and F.R. Aussenegg. (1989). *Chem. Phys. Lett.*, Vol. 159, p. 231

Druffel, E.R.M., Bauer, J.E., (2000). Radiocarbon distributions in Southern Ocean dissolved and particulate organic matter, *Geophys. Res. Lett.*, Vol. 27, pp. 1495–1498

Frew, N.M., Nelson, R.K. (1992). Scaling of marine microlayer film surface pressure-area isotherms using chemical attributes. *J. Geophys. Res.*, 97, pp. 5291–5300

Fromhertz, P. and G. Reinbold. (1988). *Thin Solid Films*, Vol. 160, p. 347

Hunter, K.A., Liss, P.S., (1981). Organic sea surface films. In: Duursma, E.K., Dawson, R. (Eds.), *Marine Organic Chemistry*. Elsevier, New York, pp. 259–298.

Ivanov, G.R. (1992). First observation of fluorescence self-quenching in Langmuir films. *Chem. Phys. Lett.*, Vol. 193, p. 323

Kostadinov, K. (2010) Robot Technology in Hybrid Assembly Approach for Precise Manufacturing of Microproducts, In: *12th Mechatronics Forum Biennial International Conference*, Book 1, pp. 293-300, Swiss Federal Institute of Technology, Switzerland: ETH Zurich.

Lancet, D. and I. Pecht. (1977). *Biochemistry*, Vol. 16, p. 5150

Lin, S. and W.S. Struve. (1991). *Photochem. Photobiol.*, Vol. 54, p. 361

Macdonald, R.I. (1990). *J. Biolog. Chem.*, Vol. 265, p. 13533

Monti, J.A., S.T. Christian, and WA. Shaw, (1978). *J. Lipid Research*, Vol.19 p. 222

Morris, S.J., D. Bradley, and R. Blumenthal. (1985). *Biochim. Biophis. Acta*, Vol. 818, p. 365

Mukherjee, S., A. Chattopadhyay, A. Samanta and T. Soujanya. (1994). *Phys. Chem.*, Vol. 98, p. 2809

Nikolelis, D.P, J.D. Brennan, R.S. Brown, and U.J. Krull. (1992). *Anal. Chim. Acta.*, Vol. 257, p. 49

Oida, T., T. Sako, and A. Kususmi. (1993). *Biophys. J.*, Vol. 64, p. 676

Paprica, P. A., N. C. Baird and N. O. Petersen. (1993). *J. Photochem. Photobiol. A: Chem.*, Vol. 70, p. 51

Peters, R. and K. Beck. (1983). *Proc. Natl. Acad. Sci. US.*, Vol. 80, p. 7183

Pogorzelski S.J. and Kogut, A. D., (2003) Structural and thermodynamic signatures of marine microlayer surfactant films, *J. Sea Research.*, Vol. 49 pp. 347–356

Pogorzelski, S.J., (2001). Structural and thermodynamic characteristics of natural marine films derived from force-area studies. *Colloids Surfaces A: Physicochem. Eng. Aspects.*, Vol. 189, 163–176

Pogorzelski, S.J., Kogut, A.D., (2001). Static and dynamic properties of surfactant films on natural waters. *Oceanologia*, Vol. 43, 223–246

Pogorzelski, S.J., Stortini, A.M., Loglio, G. (1994). Natural surface film studies in shallow coastal waters of the Baltic and Mediterranean Seas. *Cont. Shelf Res.*, Vol. 14, pp. 1621–1643

Rajarathnam, K., J. Hochman, M. Schindler, and S. Ferguson-Miller. (1989). *Biochemistry*, Vol. 28, p. 3168

Shrive, J.D.A., J.A. Brennan, R.S. Brown, and U.J. Krull. (1995). *Appl. Spectroscopy*, Vol. 49, p. 304

Suzuki, H. and H. Hiratsuka, (1988). In: *Nonlinear Optical Properties of Organic Materials*, SPIE Vol. 971, p. 97

Thompson, N. X., H. M. McConnell and T. P. Burghardt, (1984). *Biophys. J.*, Vol. 46, p. 739

Van Vleet, E.S., Williams, P.M. (1983). Surface potential and film pressure measurements in seawater systems. *Limnol. Oceanogr.*, Vol. 28, pp. 401–414

Part 4

Biodiversity and Environmental Change

7

Effect of Metal Contamination on the Genetic Diversity of *Deschampsia cespitosa* Populations from Northern Ontario: An Application of ISSR and Microsatellite Markers

Sophie Gervais and Kabwe Nkongolo
Laurentian University, Sudbury, Ontario, Canada

1. Introduction

Sudbury, Ontario (Canada) has been subjected to intense sulphur dioxide fumigation, soil contamination by aerial metal fallout and acid precipitation since the discovery of nickel and copper deposits in the late 1800s (Cox and Hutchinson, 1979; Bush and Barrett, 1993). The discovery of silver ore in Cobalt, Ontario (Canada) in the early 1900s, has been associated with arsenide and sulfarsenide mineral contamination of the soil in the region (Dumaresq, 1993). Although both these sites have been able to recover to some extent over the past 30 years due to emission reductions and remediation efforts (Dudka et al., 1995; Nkongolo et al., 2008), fact remains that such highly contaminated mine tailings often have metal concentrations that are increased to a level that are toxic for the majority of plants (Jiménez-Ambriz et al., 2007). The toxic metal pollutants, accompanied by the detrimental physical disturbances in the environment can influence plant survivorship, recruitment, reproductive success, mutation rates and migration, all of which affect the genetic diversity of the exposed populations (Deng et al., 2007). To date, hundreds of metal-tolerant genotypes have been identified from a wide variety of plant species surviving on such metal contaminated soils with many different life stories, pollinating systems and life-spans at an unexpectedly high rate (Mengoni et al., 2000; Wu et al., 1975). This type of rapid and widespread adaptation to metal pollution suggests that the evolution of metal tolerance is one of the major strategies for plant colonization of mining spoils (Deng et al., 2007).

Several investigations on the genetic diversity among metal tolerant populations relative to their non-metal tolerant counterparts have been carried out. Despite founder effect and selection, studies on *Silene paradoxa* (Mengoni et al., 2000) and *Agrostis stolonifera* (Wu et al., 1975) demonstrated that the genetic diversity of the contaminated population was the same as that of the uncontaminated populations. Other studies reported a high heterozygosity in tolerant plants such as European beech (Muller-Starck, 1985), Scots pine (Geburek et al., 1987), trembling aspen and red maple (Berrang et al., 1986). Contrary to these results, a reduction of genetic diversity has been found in other tolerant populations such as *Armeria maritime* (Vekemans and Lefèbvre, 1997).

In regards to *Deschampsia cespitosa*, two studies have examined the level of genetic diversity of populations growing on metal contaminated soils relative to those growing on uncontaminated soils. In their study, Bush and Barrett (1993) examined variation at 19 putative isozyme loci. Their population samples included individuals from 8 uncontaminated habitats, 5 from mine sites around Sudbury, Ontario and another 5 from Cobalt, Ontario. Results demonstrated lower levels of diversity in both Sudbury and Cobalt population relative to uncontaminated sites. In a subsequent study, Nkongolo et al. (2001) examined the metal contamination of the soil and associated genetic variation of *D. cespitosa* populations from Sudbury and Cobalt based on RAPD molecular markers. Results showed that the *D. cespitosa* plant has metal tolerant capabilities which allow it to survive and thrive on these mine tailings containing elevated levels of copper, nickel cobalt, aluminum, cobalt, nickel and arsenic. Although a high level of aneuploidy was observed, no link between genetic characteristics and metal contamination was made.

The methods used to infer the level of genetic diversity in these previous studies present certain disadvantages. Isozyme analysis reflects alterations in the DNA sequence through changes in amino acid composition. These changes will often alter the protein charge thereby producing a change in electrophoretic mobility which is useful in evaluating level of variation between individuals and populations on the basis of gene loci coding for specific enzymes (Hamrick, 1989). However, not all alteration in the DNA sequence will produce changes in amino acid composition therefore certain amino acid substitutions may also go undetected. Also, the limited number of isozymes (about 30) reflects only a small portion of the coding genome. The main weakness of RADP analysis is its poor reproducibility which can be caused by its sensitivity to reaction conditions such as the concentration of MgCl, *Taq* polymerase and dNTP (Qian et al., 2001). Since its introduction in 1994, the Inter-Simple Sequence Repeat (ISSR) random marker system has grown in popularity and has superseded RAPD method (Zietkiewicz et al., 1994). The ISSR marker system is based on the use of a 15-20 bp primers designed to be complimentary to microsatellite sequences found throughout the eukaryotic genome, therefore providing information at a number of different loci. Also, as a result of the longer lengths of ISSR primers in comparison to RAPD primers, the required annealing temperatures are higher and as such, non-specific binding is reduced and banding patterns have higher reproducibility (Qian et al. 2001).

Microsatellite markers have emerged as the genetic marker of choice over the last decade (Buschiazzo and Gemmell, 2006). These single-locus, hypervariable and abundant markers characterized by a Mendelian mode of inheritance, co-dominant nature, high reproducibility and easy application are a very powerful tool for assessing population genetic parameters and DNA profiling (Gaitán-Solís et al., 2002; Jones et al., 2001). Studies have reported that microsatellites are more variable than RFLPs or RAPDs and therefore have the potential to show polymorphism in species otherwise characterized by low levels of genetic diversity.

Although the information revealed by microsatellites regarding the genetic parameters of a population is highly informative, the development of these markers *de novo* is extremely expensive, labour intensive and time consuming (Saha et al., 2006). Several studies have examined the possibility of cross-species transferability of microsatellite primers pairs developed in closely related species and have reported the conservation of some loci across genera and sometimes even families (Saha et al., 2006). This method considerably reduces primer development costs while providing useful information for comparative linkage relationships between species.

The main objectives of the present study were to determine any association between metal content in the soil and the genetic variation in *D. cespitosa* populations growing in Northern Ontario using ISSR and microsatellite markers.

2. Materials and method

2.1 Sampling sites

Soil samples and leaves of *Deschampsia cespitosa* were collected from a total of nine sites located within three regions of Northern Ontario exhibiting various concentrations of metal contamination in the soil, including Sudbury, Cobalt and Manitoulin Island (Figure 1, Table 1). The sampling sites from the Sudbury region were located in Coniston, Copper Cliff, Falconbridge and Walden. The sampling sites from the Cobalt region were located on three different abandoned mine sites designated Cobalt-3, Cobalt-4 and Cobalt-5 (Figure 1). The two sampling sites from the Manitoulin Island region were located in Little Current and at the Mississagi Lighthouse. These nine sampling sites were the same as those previously characterized by Nkongolo et al. (2001) as having significant accumulation of several metals within the soil. Cobalt-4 was the most contaminated site whereas the two sampling sites from the Manitoulin Island region were the least contaminated, thus serving as control populations in this study. A total of 40 individuals were collected from each sampling area, representing approximately 5 to 20% of the entire plant population observed on site.

2.2 Metal analysis

Soil samples were analyzed in collaboration with TESTMARK Laboratories Ltd. Sudbury, Ontario, Canada. The laboratory is ISO/IEC 17025 certified, a member of the Canadian Council of Independent Laboratory (CCIL) and the Canadian Association of Environmental Analytical Laboratories (CAEAL), and is accredited by the Standards Council of Canada (SSC). The laboratory employs standard QA/QC procedures, involving blank and replicate analyses and with recovery rate of 98 ± 5% in analyses of spiked samples depending on element selected, in their inductively coupled plasma mass spectrometry (ICPMS) analyses reported here. The minimum detection limits (MDL) following microwave digestion of plant tissue Aqua Regia for elements reported here, were: Aluminum 0.05 µg/g (0.5 µg/g), Arsenic 0.05 µg/g (0.5 µg/g), Cadmium 0.05 µg/g (0.5 µg/g), Cobalt 0.05 µg/g (0.5 µg/g), Copper 0.05 µg/g (0.5 µg/g), Iron 1.0 µg/g (10 µg/g), Lead 0.05 µg/g (0.5 µg/g), Magnesium 0.2 µg/g (2.0 µg/g), Manganese 0.05 µg/g (0.5 µg/g), Nickel 0.05 µg/g (0.5 µg/g) and Zinc 0.05 µg/g (0.5 µg/g). These MDLs reflect actual sample weights and dilutions; instrument detection limits were lower.

The data for the metal levels in soil and tissue samples were analyzed using SPSS 7.5 for Windows. All the data were transformed using a $\log_{10}$ transformation to achieve a normal distribution. Variance-ratio test was done which make certain assumptions about the underlying population distributions of the data on which they are used; for example that they are normal. If the assumptions of the parametric test were violated, nonparametric test was used in place of parametric test. Kruskal-Wallis test the non-parametric analog of a one-way ANOVA was used to compare independent samples, and tests the hypothesis that several populations have the same continuous distribution. ANOVA followed by Tukey's HSD multiple comparison analysis were performed to determine significant differences ($p < 0.05$) among the sites.

2.3 DNA extraction

The total cellular DNA from individual samples was extracted from seedling tissue using the method described by Nkongolo (1999), with some modifications. The modification involved addition of PVP (polyvinylpyrrolidone) and β-mercaptoethanol to the CTAB extraction buffer.The DNA concentration was determined using the fluorochrome Hoechst 33258 (bisbensimide) fluorescent DNA quantitation kit from Bio-Rad (cat. # 170-2480) and the purity was determined using a spectrophotometer (Varian Cary 100 UV-VIS spectrophotometer).

2.4 ISSR analysis

All DNA samples were primed with each of the nine primers (ISSR 17898B, UBC 818, UBC 823, UBC 825, UBC 827, UBC 835, UBC 841, UBC 844, and UBC 849) (Mehes et al., 2007). The ISSR amplification was carried out in accordance with the method described by Nagaoka and Ogihara (1997), with some modifications described by Mehes et al. (2007). All PCR products were loaded into 2% agarose gel in 1X Tris-Borate-EDTA (TBE) buffer. Gels were pre-stained with 4μl of ethidium bromide and run at 3.14V/cm for approximately 120 minutes. These agarose gels were visualized under UV light source, documented with the Bio-Rad ChemiDoc XRS system and analyzed for band presence or absence with the Discovery Series Quantity One 1D Analysis Software.

ISSR assays of each population were performed at least twice. Only reproducible amplified fragments were scored. For each sample, the presence or absence of fragments was recorded as 1 or 0, respectively and treated as a discrete character. Pairwise comparison of banding patterns was evaluated using RAPDistance, version 1.04 (Armstrong et al. 1994). The data were analyzed to generate Jaccard's similarity coefficients and genetic distances. These similarity coefficients were used to construct dendrograms, using neighbour-joining analysis (Saitou and Nei 1987). Analysis of molecular variance (AMOVA) was applied, to estimate variance components for ISSR phenotypes. Variations were partitioned among individuals (within regions) and between regions. Levels of significance were determined using nonparametric permutational methods with the Winamova program (Excoffier et al., 1992).

2.5 Microsatellite analysis

A total of 31 microsatellite primers, developed in several members of the *Poaceae* family (Table 2), were screened for cross-species consrvation in *Deschampsia cespitosa*. Of these, 5 were from *Elymus caninus* (Sun et al., 1999), 7 were from *Avena sativa* (Li et al., 2000), 3 were from *Triticum aestivum* (Röder et al., 1995) and 17 were developed in *Hordeum vulgare* (Liu et al., 1996; MacRitchie and Sun, 2004; Struss et al., 1998). The microsatellites primers used in this study were selected based on the phylogenetic relationship between the species of origin and *D. cespitosa*, the polymorphic index of the alleles within their respective species and, in some cases, previous reports of cross-species conservation of the microsatellite locus. DNA amplification was performed following the procedure described by Mehes et al. (2010). Of the 31 microsatellite primer pairs screened, only those that successfully amplified a clear, reproducible, distinguishable band, demonstrated microsatellite characteristics and showed a certain degree of polymorphism were used in the study. The agarose gels were documented using the Bio-Rad ChemiDoc XRS system and analyzed with the Discovery Series Quantity One 1 D Analysis Software. Nine microliters of 3X loading buffer (10 mM NaOH, 95% formamide, 0.05% bromophenol blue, 0.05% xylene cyanol) was added to the remainder of the 18 µl of the amplified products and 2.5 µl of 3X loading buffer was added to the mixture of 5 µl of water and 1.5 µl of 10 bp ladder (Invitrogen). The samples and the 10 bp ladder (Invitrogen) were denatured at 99°C for 10 min and snap cooled for 2 min on ice prior to loading on denaturing gels. The PCR products were electrophoresed on a 0.4 mm denaturing 6% polyacrylamide gels containing 8 M Urea and 1X TBE buffer at constant power of 73 W, 2 100 V and 90 mA and were equilibrated to 55°C (DNA Sequencing System, FisherBiotech, Fisher Scientific). The amplifications products were visualized with the Silver Staining Sequence DNA Sequencing System according to the manufacturers protocol (Promega Corporation). Resolved fragments were sized by The Quantity One and Genescan softwares.

Fig. 1. Map of Northern Ontario illustrating the three regions, Manitoulin Islands, Sudbury and Cobalt, where samples of *Deschampsia cespitosa* were collected for this study.

The presence and absence of alleles yielded by the microsatellite primer pairs were scored as 1 or 0, respectively in order to determine the polymorphism of each locus. Such designations were carried out with the Quantity One software by establishing the alleles of interest and comparing them to the 10 bp ladder which served as a marker system. The data was computed into the Popgene software, version 1.32 (Yeh et al., 1997) and used to determine the intra- and interpopulation genetic diversity parameters such as number of alleles per locus, mean number of alleles across loci, percentage of polymorphic loci and Shannon's information index. In order to determine whether the observed allelic proportions met Hardy-Weinberg expectations, the populations were tested using an exact test (Guo and Thompson, 1992) by the computer program GENEPOP version 1.2 (Raymond and Rousset,

1995). Hardy-Weinberg Equilibrium deviations were further tested with heterozygote deficiency and excess alternative hypotheses. Wright's *F* statistics including inbreeding coefficients and fixation index (Weir and Cockerham, 1984) as well as gene flow were determined using Popgene. The PowerMarker software, version 3.25 (Liu and Muse, 2005) was used to calculate the genetic distances between the populations based on the Cavalli-Sforza and Edwards Chord's Distance (1967). Finally, the relationship between the matrices based on microsatellite data and geographical location was calculated using the Pearson's correlation coefficient and the significance of the correlation was determined by the Mantel test (Mantel, 1967) using XLSTAT version 7.5 software package (www.xlstat.com).

Region	Site	Latitude	Longitude	Altitude
Manitoulin Island	Little Current	N°45°58.755′	W°81° 54.950′	180m
	Mississagi Lighthouse	N°45°53.493′	W°83° 13.552′	176m
Sudbury	Coniston	N°46°28.703′	W°80° 50.862′	252m
	Copper Cliff	N°46°30.246′	W°81° 02.819′	292m
	Falconbridge	N°46°34.619′	W°80° 48.701′	323m
	Walden	N°46°25.842′	W°81° 04.627′	260m
Cobalt	Cobalt-1	N°47°22.524′	W°79°41.008′	313m
	Cobalt-2	N°47°23.206′	W°79° 39.683′	302m
	Cobalt-3	N°47°21.725′	W°79° 38.861′	347m

Table 1. Geographic coordinates of the nine *D. cespitosa* sampling sites located throughout Northern Ontario.

Species of origin	Locus	Primer Sequence (5′→3′)	Repeat	Number of alleles	Tm (°C)	Expected size (bp)
Elymus caninus	ECGA22	gaaggtgactaggtccaac atagtctcggtcaggctc	$(CT)_{27}$	13	54	166
	ECGA125	tgcttccaacttgctca tgcatctgtgtgtccaca	$(AG)_{23}$	10	54	204
	ECGA126	gtcactagtggatcgtgcc gatttggtgtcgttctgatc	$(GA)_{15}$	9	54	186
	ECGA114	cttatatcttgtgggttatcat gatctgatacgtgacatctca	$(TC)_{15}$	8	54	129
	ECGA210	cgacaactagtggatcaaa gaagtactctcgagaagctt	$(GA)_{22}$	6	54	196
Avena sativa	AM1	ggatcctccagcctgttga ctcatccgtatgggcttta	$(AG)_{21}(CAGAG)_6$	5	46	204
	AM3	ctggtcatcctcgccgttca catttagccaggttgccaggtc	$(AG)_{35}$	5	51	280
	AM4	ggtaaggtttcgaagagcaaag gggctatatccatccctcac	$(AG)_{34}$	6	48	166
	AM14	gtggtgggcacggtatca tgggtggcgaagcgaatc	$(AC)_{21}$	4	48	133

	AM22	attgtatttgtagccccagttc aagagcgacccagttgtatg	$(AC)_{22}$	8	46	138
	AM30	tgaagatagccatgaggaac gtgcaaattgagtttcacg	$(GAA)_{14}$	6	43	203
	AM31	gcaaaggccatatggtgagaa cataggtttgccattcgtggt	$(GAA)_{23}$	6	47	186
Triticum aestivum	WMS2	ctgcaagcctgtgatcaact cattctcaaatgatcgaaca	$(CA)_{18}$	4	50	132
	WMS6	cgtatcacctcctagctaaactag agccttatcatgaccctacctt	$(GA)_{40}$	4	50	205
	WMS30	atcttagcatagaagggagtggg ttctgcaccctgcctgat	$(AT)_{19}(GT)_{15}$	7	60	206
Hordeum vulgare	GMS005	actacgtccagtcgtttcc tgaacaccacgggttcatc	$(GT)_{19}T(CT)_2(GT)_{13}$	10	55	170
	GMS006	tgaccagtaggggcagtttc ttcttctccctcccccac	$(GA)_2ATA(GA)_{19}$	14	55	154
	GMS046	atgtatttatcacccacccagc aaggcattagaaccggcac	$(GA)_{13}$	15	60	156
	GMS056	gagaaacgcagctgtggc gtcaccgaggccttcctc	$(GA)_{11}$	13	60	137
	GMS114	aaccagtgggttttaacccc tgccaccacatgcatacac	$(GT)_{11}$	11	55	152
	HVM3	acaccttcccaggacaatccattg agcacgcagagcaccgaaaaagtc	$(AT)_{29}$	3*	55	188
	HVM5	aacgacgtcgccacacac aggaacgaagggagtattaagcag	$(GT)_6(AT)_{16}$	4**	55	202
	HVM7	atgtagcggaaaaaataccatcat cctagctagttcgtgagctacctg	$(AT)_7$	2**	55	174
	HVM20	ctccacgaatctctgcacaa caccgcctcctctttcac	$(GA)_{19}$	5*	T1	151
Hordeum vulgare	HVM27	ggtcggttcccggtagtg tcctgatccagagccacc	$(GA)_{14}$	+	T1	192
	HVM40	cgattcccttttcccac attctccgccgtccactc	$(GA)_6(GT)_4(GA)_7$	3**	T1	160
	HVM44	aaatctcaggttcgtgggca ccacggagaccacctcactt	$(GA)_8$	4*	T1	114
	HVM51	tctaaattaccttcccagcca aagcagacatgtaggaggtca	$(GA)_3(GGGA)_3(GA)_8$	4*	T1	151
	HVM60	caatgatgcggtgaactttg cctcggatctatgggcctt	$(AG)_{11}(GA)_{14}$	3*	T1	115
	HVM65	agacatccaaaaaatgaacca tggtaacttgtcccccaaag	$(GA)_{10}$	5*	T1	129
	HVM 68	aggaccggatgttcataacg caaatcttccagcgaggct	$(GA)_{22}$	+	T1	204

*: Number of alleles expected in the *Avena* species
**: Number of alleles expected in the *Elymus* species
+: Polymorphic in four crossed mapping populations of barley
T1: Touchdown PCR with 18 cycles 64°C-55°C and 30 cycles at 55°C (Liu et al., 1996)

Table 2. List of the microsatellite primer pairs screened for transferability in *D. cespitosa*.

3. Results

3.1 Heavy metal analysis

Recovery and precision for all elements in reference soil samples were within acceptable range. The estimate levels of metal content in different sites are illustrated in Figure 2. The levels of the metals measured were low in the control sites. Overall, the results indicated that nickel and copper continue to be the main contaminants of soil in Coniston while cadmium, cobalt, copper, lead, zinc and to some extent nickel, were found in high concentration in Cobalt sites. The Cobalt 4 site is by far the most contaminated of all the sites (Fig. 2). For example, the average mean level of zinc at Cobalt 4 is at least twenty-one fold than that of Sites from Sudbury. Cobalt-4 was also among the sites with the highest accumulation of copper, lead, and nickel.

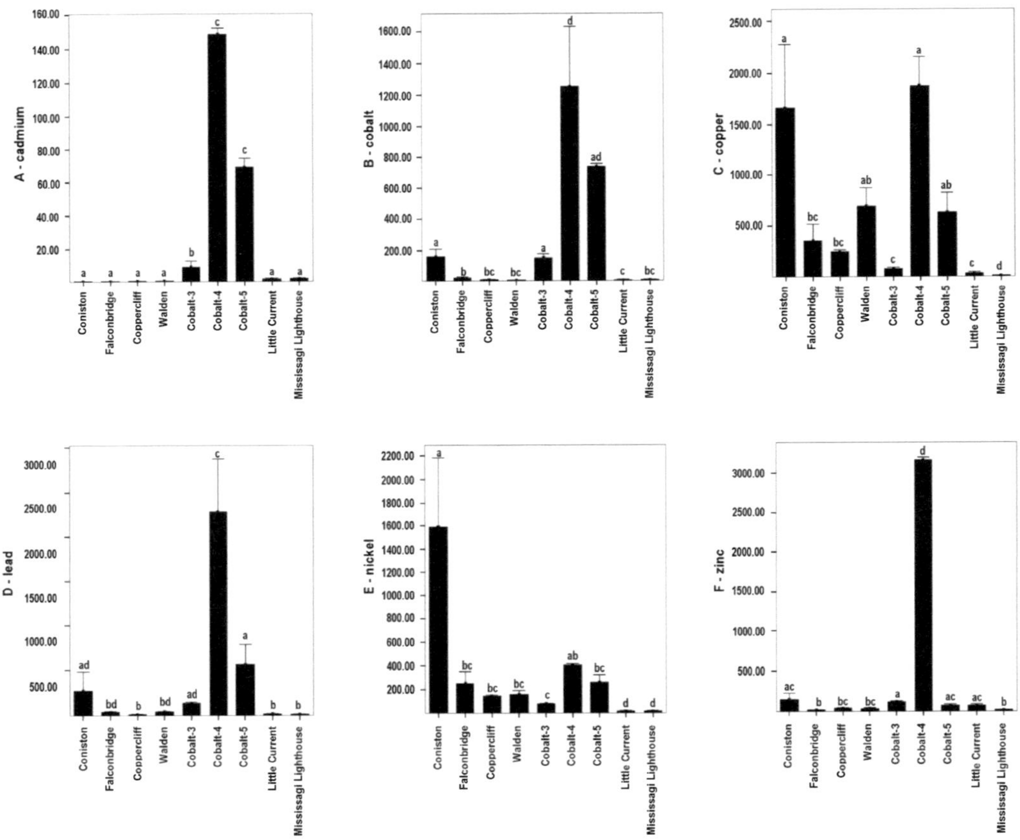

Fig. 2. Cadmium, cobalt, copper, lead, nickel, and zinc concentrations in soil samples collected from the Sudbury, Cobalt, and Manitoulin Island regions. Means with common notations are not significantly different as indicated by Tukey HSD analysis ($p > 0.05$).

Unlike the Sudbury sites, which are located near smelters, Cobalt-4 site is located in the concrete remains of the foundation of an abandoned mine site. This could explain the extraordinarily high heavy metal accumulation in that area. This particular site also does not appear to have been detoxified or rehabilitated like the Sudbury sites have been. Cobalt 3 showed relatively lower level of heavy metal accumulation than Cobalt 4 and 5 and was typically in the middle of the spectrum of contaminated sites. The two control sites, Little Current and Mississagi Lighthouse were always among the least contaminated for the metals analysis. Three of the Sudbury sites, Falconbridge, Coppercliff, and Walden were also consistently among the least contaminated. Coniston was found to be on average significantly more contaminated than the other Sudbury sites (Fig. 2).

3.2 ISSR analysis

Five of the nine primers screened (Table 3) produced good amplification products ranging from 160 bp to 1300 bp. They include 17898B, UBC 818, UBC 827, UBC835 and UBC 841 had good amplifications products (Figure 8). The level of polymorphic loci among populations was 63 %. This value was much lower than the polymorphisms level (90%) reported in RAPD analysis of the samples from the same sites (Nkongolo et al., 2001). The polymorphism within each population varied between 44% observed in Cobalt-5 to 92% in Falconbridge (Table 4). The overall values for the three regions were compared. The highest level of polymorphism was observed in samples from the Sudbury region (Coniston, Falconbridge, Copper Cliff, Walden) with an average of 74 %, followed by the Manitoulin Island region (Little Current and Mississagi Lighthouse) with an average of 69%. The lowest level of polymorphic loci was observed in samples originating from the Cobalt region (Cobalt-3, Cobalt-4, and Cobalt-5) with a mean value of 46%. Interestingly, the Cobalt region was also the site which exhibited the highest accumulation of metals in the soil.

Primer Identification	Nucleotide sequence (5′→ 3′)	Amplification	Fragment size range (bp)
ISSR 17898B	CACACACACACAGT	Good	200-1300
UBC 818	CACACACACACACACAG	Good	400-1100
UBC 823	TCTCTCTCTCTCTCTCC	Absent	-----
UBC 825	ACACACACACACACACT	Fair	650-1000
UBC 827	ACACACACACACACACG	Good	400-1000
UBC 835	AGAGAGAGAGAGAGAGYC	Good	200-1000
UBC 841	GAAGGAGAGAGAGAGAYC	Good	160-850
UBC 844	CTCTCTCTCTCTCTCTRC	Fair	500-650
UBC 849	GTGTGTGTGTGTGTGTYA	Fair	300-500

Table 3. Nucleotide sequences of the primers used to produce ISSR profiles by amplification of genomic DNA from nine populations of *Deschampsia cespitosa*.

Region	Site	Total number of bands	Number of polymorphic bands	Polymorphic bands (%)	Mean polymorphism per region (%)
Sudbury	Coniston	75	54	72	74 (Sudbury)
	Falconbridge	72	66	92	
	Copper Cliff	60	40	67	
	Walden	72	47	65	
Cobalt	Cobalt-3	75	36	48	46 (Cobalt)
	Cobalt-4	69	32	46	
	Cobalt-5	81	36	44	
Manitoulin Island	Little Current	66	46	70	69 (Manitoulin)
	Mississagi Lighthouse	87	59	68	

Table 4. Levels of polymorphisms within *Deschampsia cespitosa* populations from Northern Ontario generated with ISSR primers.

The between- populations variance contributed 13.6 % of the total variance while the within-population variance accounted for 71.2%. Using a nonparametric test, we found that between-population differences were significant (Table 5). No single locus appears to be specific to contaminated sites.

Source of variation	df	MS	Variance Component	% Total	P
Among regions	2	1.224	0.034	13.60	0.001
Populations within region	6	0.562	0.053	15.23	0.001
Individuals within populations	63	0.277	0.281	71.17	0.001

Table 5. Analysis of molecular variance (AMOVA) for ISSR variation among *Deschampsia cespitosa* populations from several locations in Northern Ontario.

In general, the genetic distances among the nine *D. cespitosa* populations from Northern Ontario values varied from 0.059 to 0.488. The scale utilized denotes a 0 for identical populations to a 1 for populations that are different for all criteria (Table 6). The closest genetic distance value (0.06) was observed between the populations from Cobalt-5 and Cobalt-3. The two most genetically distant populations were Cobalt-5 and Falconbridge (0.49). The genetic distance data also showed that the four *D. cespitosa* populations from the Sudbury area (Coniston, Falconbridge, Copper Cliff and Walden) were closely related. These data also revealed that the *D. cespitosa* populations from the Cobalt region (Cobalt-3, Cobalt-4, and Cobalt-5) were closely related to the *D. cespitosa* population from Little Current (Table 6, Fig. 3). The results were supported by the cluster analysis that illustrated that the four *D. cespitosa* populations from the Sudbury region clustered together and the three Cobalt populations clustered with the populations from Little Current and Mississagi Lighthouse (Fig. 3). Overall, the molecular analysis using ISSR markers showed that the *D. cespitosa* populations from Northern Ontario are different but closely related.

	1	2	3	4	5	6	7	8	9
1	0.000	0.147	0.189	0.1944	0.366	0.391	0.375	0.308	0.400
2		0.000	0.222	0.278	0.476	0.452	0.488**	0.425	0.395
3			0.000	0.114	0.341	0.317	0.390	0.325	0.333
4				0.000	0.308	0.282	0.359	0.289	0.297
5					0.000	0.184	0.059*	0.189	0.333
6						0.000	0.237	0.211	0.350
7							0.000	0.194	0.385
8								0.000	0.222
9									0.000

1 represents D. cespitosa population from Coniston; 2: D. cespitosa population from Falconbridge; 3: D. cespitosa population from Copper Cliff; 4: D. cespitosa population from Walden; 5: D. cespitosa population from Cobalt-3; 6: D. cespitosa population from Cobalt-4; 7: D. cespitosa population from Cobalt-5; 8: D. cespitosa population from Little Current; 9: D. cespitosa population from Mississagi Lighthouse.

Table 6. Distance matrix generated using the neighbor-joining analysis from *Deschampsia cespitosa* ISSR data (RAPDistance version 1.04).

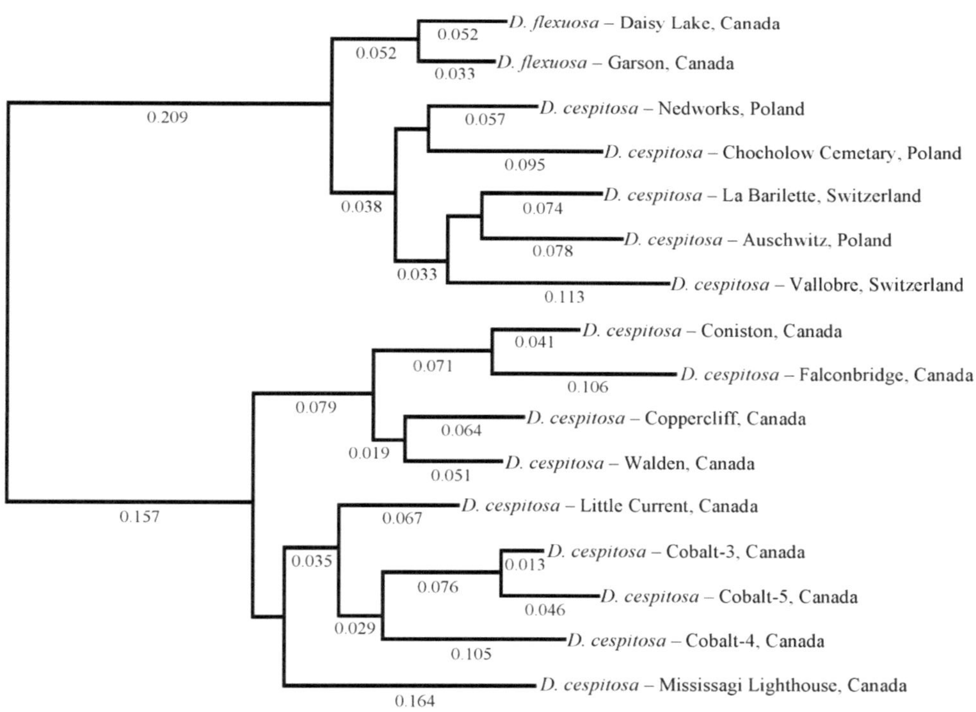

Fig. 3. Dendrogram of the genetic relationships between the nine *D. cespitosa* populations from Northern Ontario using data generated from the Jaccard similarity matrix from ISSR profiles. The values below the branches indicate the patristic distances based on the neighbor-joining analysis. The *D. cespitosa* populations from Europe and the *D. flexuosa* populations (Northern Ontario) were used as outgroups.

3.3 Nuclear microsatellite analysis

Only primer pairs HVM3, HVM5, HVM20, HVM65 and WMS6 successfully amplified a clear, reproducible, distinguishable band within an acceptable range of the expected size while demonstrating polymorphism. This represents only 14% of the microsatellite primer pairs screened for transferability within *D. cespitosa*. All alleles detected were scored according to the guidelines previously outlined. Every possible microsatellite loci pair, in every population, was surveyed for association following the null hypothesis that genotypes at one locus are independent from genotypes at the other locus. Based on the results, no evidence of linkage disequilibrium between the microsatellite loci used in this study was observed. Only data related to genetic diversity are discussed in the present report.

3.3.1 Genetic diversity

Genetic diversity within each individual, population, region and locus was assessed using standard descriptive statistics. The five polymorphic microsatellite markers detected a total of 40 alleles. The mean number of alleles per locus across populations was 2.1, 2, 7.3, 4.6 and 2.6 for locus HVM 3, HVM 5, HVM 20, HVM 65 and WMS 6 respectively. The mean number of alleles per populations across loci was 3 for Little Current, 3.2 for Mississagi Lighthouse and Walden, 3.6 for Falconbridge and Copper Cliff, 3.8 for Coniston and Cobalt-5, 4 for Cobalt-4 and 5.2 for Cobalt-3. At the population level, the observed mean heterozygosity (H_O) and the expected mean heterozygosity (H_E) ranged from 0.413 and 0.48 in the Mississagi Lighthouse population to 0.65 and 0.76 in the Cobalt-3 population. At the regional level, the H_O and the H_E ranged from 0.40 and 0.46 in the Cobalt region to 0.34 and 0.42 in the Manitoulin Island region. Finally, the H_O and the H_E observed by individual loci ranged from 0.28 and 0.44 for HVM3 to 0.85 and 0.99 for HVM20. Shannon's Information Index (i) was also calculated as an additional measure of genetic diversity. Values varied from 0.54 at the HVM5 locus to 2.58 at the HVM20 locus with a mean of 1.417 (Table 7).

Inbreeding is defined as the non-random uniting of gametes which results in a decrease of heterozygotes. The level of inbreeding within a population is determined by the inbreeding coefficient, F_{IS}, where -1 (all individuals heterozygous) $\leq F_{IS} \leq 1$ (no observed heterozygotes). The inbreeding coefficients (F_{IS}) were determined for each population per loci based on the null hypothesis of no inbreeding represented as F_{IS} = -1. All nine populations exhibited a negative F_{IS} value at loci HVM3, HVM5 and HVM20 indicating an excess of heterozygotes. Only Walden and Cobalt-3 populations presented negative F_{IS} values at locus HVM65. Finally, six of the nine populations exhibited negative F_{IS} values at the WMS6 locus. As a result, the overall inbreeding coefficient for *D. cespitosa* populations were -0.18, -0.08, -0.07, -0.35, -0.19, -0.07, -0.12, -0.21 and -0.17 for Coniston, Falconbridge, Copper Cliff, Walden, Cobalt-3, Cobalt-4, Cobalt-5, Little Current and Mississagi Lighthouse, respectively.

The fixation index (F_{ST}) is a measure of the extent of genetic differentiation among populations due to genetic drift. Values can range from 0.0, indicating no differentiation, to 1.0, indicating complete differentiation. However, because the observed maximum is usually much less than 1.0, a value between 0.0 and 0.05 is considered as little genetic differentiation, 0.05 and 0.15 as moderate genetic differentiation, 0.15 and 0.25 as great genetic differentiation and values above 0.25 as very great genetic differentiation. The F_{ST} values were 0.079 for HVM5 and 0.088 for HVM20, indicating moderate genetic differentiation. Locus HVM65 and locus WMS6 demonstrated great genetic differentiation with 0.22 and 0.22 F_{ST} values, respectively (Table 8). Finally, HVM3 locus represented a very great deal of genetic differentiation with an F_{ST} value of 0.311 (Table 8).

The F_{ST} values were subsequently used to estimate the level of gene flow (N_m) for each locus according to Nei (1987), where $N_m = 0.25(1-F_{ST})/F_{ST}$. The mean level of gene flow was 0.933 with individual N_m values ranging from 0.5424 for the HVM65 locus to 2.347 for the HVM5 locus (Table 8). Genetic distance coefficients were calculated according to the Cavalli-Sforza and Edwards chord distance (D_C). This particular scale ranges from 0, indicating no genetic difference to 1, indicating differences at all conditions criteria. The genetic distance coefficients varied between 0.57 between the Coniston and Cobalt-3 populations and 0.18 between the Falconbridge and Copper Cliff populations (Table 9). Based on these values, an un-rooted Neighbour-Joining phylogenetic tree was constructed with 101-bootstrap. The resulting tree illustrates three major clades (Fig. 4). The first is composed of the Little Current population, the second comprises all four Sudbury region populations and the Mississagi Lighthouse population and the third includes all three Cobalt region populations. Within the second clade, the Mississagi Lighthouse population is the most distantly related, while the Falconbridge and Copper Cliff populations and the Walden and Coniston populations form clusters. Within the third clade, the Cobalt-5 population appears to be the most distantly related.

The Mantel test results reveal a significant correlation between the two distance matrices ($r = 0.514$, $p = 0.01$) suggesting a congruence between the genetic distance generated from microsatellite data and the geographical distance between populations.

Population	N_A*	H_O*	H_E*	I*	F_{IS}*
Coniston	3.8	0.680	0.582	1.0572	-0.179
Falconbridge	3.6	0.580	0.538	0.9689	-0.083
Copper Cliff	3.6	0.500	0.468	0.8506	-0.071
Walden	3.2	0.700	0.529	0.8887	-0.346
Cobalt-3	5.2	0.760	0.645	1.2141	-0.190
Cobalt-4	4.0	0.680	0.640	1.1305	-0.066
Cobalt-5	3.8	0.600	0.540	0.9720	-0.118
Little Current	3.0	0.620	0.518	0.8401	-0.210
Mississagi Lighthouse	3.2	0.480	0.413	0.7379	-0.174
Mean	3.71	0.542	0.622	0.9622	-0.142
Standard Error	± 0.05	±0.025	±0.031	±	± 0.01

*Genetic diversity descriptive statistics. NA, mean number of alleles across all loci; HO, observed heterozygosity; HE, expected heterozygosity; I, Shannon's information index; FIS, inbreeding coefficient.

Table 7. Genetic diversity estimates for Deschampsia cespitosa populations.

Locus	A	Obs. Gen	N_E	H_O	H_E	I	PIC	F_{ST}	N_m
HVM3	6	5	2.1	0.444	0.280	0.750	0.350	0.311	
HVM5	2	2	2	0.467	0.340	0.543	0.294	0.079	2.347
HVM20	17	45	7.3	0.989	0.851	2.584	0.905	0.088	
HVM65	11	24	4.6	0.478	0.666	2.056	0.824	0.219	0.542
WMS6	4	8	2.6	0.722	0.570	1.287	0.648	0.224	
Mean			3.72	0.620	0.542	1.444	0.604	0.184	0.933
Standard Error			±1.011	±0.105	±0.105	±0.865	±0.19	±0.09	

*Genetic diversity descriptive statistics. A, total number of alleles; Obs. Gen, total number of observed genotypes; NE, effective number of alleles; HO, observed heterozygosity; HE, expected heterozygosity; I, Shannon's information index; PIC, polymorphic information content; FST, fixation index (Wright 1978); Nm, gene flow estimate as per FST=0.25(1-FST)/FST (Nei 1987).

Table 8. Genetic diversity parameters for the five microsatellite primer pairs of *Deschampsia cespitosa.*

	1	2	3	4	5	6	7	8	9
1	0.0000	0.2524	0.2628	0.1836*	0.5657**	0.5145	0.5210	0.4696	0.4001
2		0.0000	0.1996	0.2630	0.5183	0.4907	0.5582	0.4720	0.4639
3			0.0000	0.3141	0.5338	0.5007	0.5486	0.4413	0.4385
4				0.0000	0.5557	0.4787	0.5227	0.4960	0.4285
5					0.0000	0.3463	0.4672	0.5261	0.5090
6						0.0000	0.4286	0.4909	0.5004
7							0.0000	0.4387	0.4053
8								0.0000	0.4292
9									0.0000

1 represents D. cespitosa population from Coniston; 2 D. cespitosa population from Falconbridge; 3 D. cespitosa population from Copper Cliff; 4 D. cespitosa population from Walden; 5 D. cespitosa population from Cobalt-3; 6 D. cespitosa population from Cobalt-4; 7 D. cespitosa population from Cobalt-5; 8 D. cespitosa population from Little Current; 9 D. cespitosa population from Mississagi Lighthouse. * Most genetically similar populations (i.e. Coniston and Walden). **Most genetically different populations (i.e. Coniston and Cobalt-3).

Table 9. Cavalli-Sforza and Edward's chord's distance matrix (1967) generated from microsatellite data used in neighbor-joining analysis of *Deschampsia cespitosa* populations.

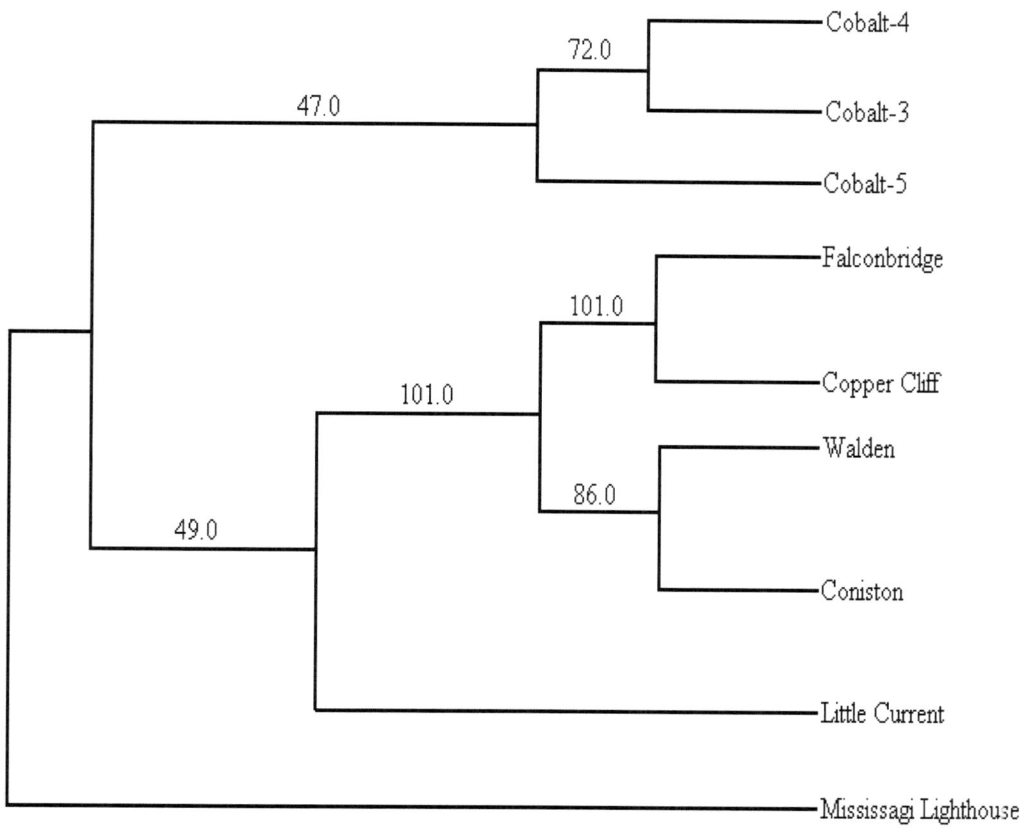

Fig. 4. Dendrogram of the genetic relationship between the nine populations of *Deschampsia cespitosa* from Northern Ontario using the data generated from Chord's distance (Cavalli-Sforza and Edwards, 1967) based on the microsatellite profiles. This is an un-rooted tree based on neighbor-joining (NJ) analysis constructed with 101-bootstrap.

4. Discussion

4.1 ISSR analysis

In general, the efficiency of a molecular marker technique depends on the amount of polymorphism it can detect among the set of accessions investigated. In the present study, the level of polymorphism detected with the ISSR system was lower than that observed with the RAPD method. Similar results were obtained by Fang and Roose (1997) who showed that RAPD PCR had identified a higher level of variation in *Citrus spp.* than ISSRs. Other studies conducted by Nagoaka and Ogihara (1997), Nkongolo et al. (2005), Raina et al. (2001) and Qian et al. (2001) have shown that ISSRs reveal a higher level of variation than RAPD markers in other plant species. Technically, RAPD and ISSR markers are different systems targeting different areas of the genome. RAPD markers cover the entire genome, revealing length polymorphism in coding or non-coding regions as well as repeated or single-copy sequences (Williams et al. 1993). Unlike RAPD products, the origin of the amplification products in ISSR-PCR is known to be from the sequences between two simple-sequence repeat primer sites where length variation does not necessarily reflect simple-sequence length polymorphism (Zietkiewicz et al. 1994). The level of variation detected with each system greatly depends on the primers used, thus making comparisons regarding levels of polymorphisms generated with RAPD and ISSR primers rather inconsistent.

The ISSR analysis revealed great variation in regards to the genetic relatedness of the samples analyzed. In general, the genetic distance values revealed that *D. cespitosa* from the nine Northern Ontario sites are genetically close. The genetic distance values between Cobalt-3 and Cobalt-5 indicate that these two populations are quite genetically similar. This suggests that these sites were likely seeded with the same genetic materials. The relative small genetic distance values among Sudbury populations and their clustering on the dendrogram are consistent with the previous RAPD data (Nkongolo et al., 2001). These findings also corroborate with the speculations from several ecologists that these populations might be the result of a single colonization event (Winterhalder, 2002; Personal communication). In general, the genetic similarity between the nine *D. cespitosa* populations from Northern Ontario may suggest that these populations could have originated from a common source. Furthermore, based on the genetic distance data, the theory of Cobalt and Little Current populations as the source of *D. cespitosa* which colonized the Sudbury area around 1972 can not be rejected.

Previous genetic analysis of these populations aiming at establishing relationships among these nine sites using RAPD markers was inconclusive (Nkongolo et al., 2001). Also, the study conducted by Bush and Barrett (1993) using isozyme markers indicated that the Sudbury and Cobalt samples showed enough variation to reject the theory of Cobalt *D. cespitosa* colonizing Sudbury. Although isozymes and ISSRs allow the analysis of genetic variability in plant species, fundamental differences exist between these two methods. Isozyme analysis reflects alterations in the DNA sequence of coding regions in the genome leading to changes in amino acid composition which can go undetected (Hamrick 1989). ISSRs target microsatellites sequences located throughout the entire eukaryotic genome, most of which are selectively neutral areas. These areas are known to evolve rapidly and as such, have been deemed good tools for any study in genetic diversity in many organisms (Blair *et al.*, 1999).

Attempts were made in the present study to use environmental conditions for appropriately interpreting genetic information. The effects of novel and toxic environments have been examined in considerable detail in the study of life history evolution. There are theoretical reasons for expecting the genetic variance of a life history character to increase when the population is challenged with a novel environment, an expectation that has been upheld empirically by numerous studies (Service and Rose, 1985; Holloway et al., 1990). If metal tolerance is controlled by many genes as suggested by Von Frenkell-Insam and Hutchinson (1993) and McNair (1993), it is very likely that allelic frequency in an out-crossing and perennial species like *D. cespitosa* will be maintained over time resulting in a neutral genetic variation. The high level of genetic variability within *D. cespitosa* populations from the Sudbury region could be ascribed in part to these conditions.

If the toxic stress continues at a sub-lethal level for many generations, resistance could develop, resulting in a decrease in genetic variation through selection. This might be the case of *D. cespitosa* populations from Cobalt where the high accumulation of heavy metal for several years appears to have significant impacts on the genetic structure of the *D.*

cespitosa populations in that region by processes which are assumed to have selective effects. Metals impose severe stress on plants, especially in the rooting zone, which has led to the evolution of metal-resistant ecotypes in several herbaceous species like *D. cespitosa* (Cox and Hutchinson, 1979). Plants possess homeostatic cellular mechanisms to regulate the concentration of metal ions inside the cell to minimize the potential damage that could result from the exposure to nonessential metal ions. These mechanisms serve to control the uptake, accumulation and detoxification of metals (Foy et al., 1978). Selection of metal-resistant genotypes has been demonstrated to occur rapidly, within one or two generations, in populations that contain the necessary genetic information (Wu et al. 1975). These authors identified two factors that may affect the plant's ability to tolerate metals; the intensity of the contamination and the amount of time the population has been exposed to the toxic levels. The populations of *D. cespitosa* in Cobalt have been there for a much greater amount of time than the populations in Sudbury and the Cobalt soil is more contaminated than the Sudbury soil. This has resulted in a decrease and possibly a loss of alleles at some loci and many rare alleles that has lead ultimately to a lower genetic diversity in *D. cespitosa* populations from the Cobalt region. Evidence of a loss of genetic diversity at the population level caused by pollution has been demonstrated in other species (Lopes et al., 2004; Van Straalen and Timmermans, 2002). The low level of genetic variation in samples from *D. cespitosa* population from Cobalt-3 could be explained by the fact that this abandoned mine waste site that was likely as contaminated as Cobalt-4 and Cobalt-5 sites has been covered with a clay cap. This fresh clay that was brought in was not as contaminated as the surrounding area.
There is also a slight possibility that the small level of polymorphic loci detected in Cobalt samples could be caused by a founder effect. This is a form of genetic bottleneck occurring where new populations are established by a small number of individuals, or by a group of individuals whose genetic variation is not representative of the parent population. However, the possibility of a founder effect occurring only in Cobalt populations and not in Little Current where the *D. cespitosa* population has been isolated for several generations is quite small (Peter Beckett, 2006, personal communication). It should also be pointed out that *D. cespitosa* is an out-crossing species which produces a lot of seeds every year. These characteristics alone usually negate possible founder effects in many species (Hedrick et al., 1976). Muller et al. (2004) also indicated that high frequency of heavy metal tolerance in natural populations can reverse the effects of an initial genetic bottleneck.

4.2 Microsatellite analysis

The various application of microsatellite markers are the direct result of their hypervariability, co-dominant nature, abundance throughout the genome and reproducibility. Their primary disadvantage lies is their time consuming and expensive *de novo* development and, as such, has been restricted to a few agriculturally important crops. In light of this, a growing number of studies have examined the ability of microsatellite primer pairs to amplify across closely related species. In this study, we were able to identify several polymorphic microsatellite loci in *Deschampsia cespitosa* via SSR primer pairs developed in *Hordeum vulgare* (barley) and *Triticum aetivum* (wheat). The five microsatellite primer pairs, namely HVM3, HVM5, HVM20, HVM65 and WMS6 were originally selected because of their high polymorphic index and documented conservation across species. Primers HVM3, HVM20 and HVM65, originally developed in barley showed conservation in *Avena* species with polymorphic information content (PIC) of 0.44, 0.77 and 0.72, respectively. In addition, a PIC of 0.15 for HVM3 and 0.50 for HVM20 was reported in oat cultivars (Li et al., 2000). The HVM5 primer originally developed from barley, showed conservation while displaying polymorphism in *Elymus* and *Pseudoroegneria* species (MacRitchie and Sun, 2004). WMS6, originally developed in wheat, had been documented as conserved and polymorphic in both barley and rye (Röder et al., 1995). In *Deschampsia cespitosa,* these primers ranged from having 2 alleles at the HVM5 locus with a PIC of 0.29 to 17 alleles at the HVM20 with PIC values of 0.91. Our findings are in accordance with other studies conducted by Gaitán-Solís et al. (2002), and Saha et al. (2004), which all report the occurrence of cross-species microsatellite primer pair transferability at an elevated rate. Such success depends heavily on the conservation of priming sites within the flanking regions and as such the evolutionary relatedness of the species sampled. Particular to the Poaceae family, believed to have radiated about 60 million years ago, genetic mapping has revealed remarkable conservation of gene content and gene order. Studies have shown that the linear organization of genes in some nine different genomes varying in chromosome number, from 5 to 12 and nuclear DNA amount, from 400 to 6000Mb, can be described in terms of only 25 rice linkage blocks (Gale and Devos, 1998). We also report shorter SSRs at HVM3, HVM5 and HVM 20 loci in *D. cespitosa* than in the species of origin with a difference in size of 36bp, 33bp and 5bp respectively. HVM65 has an allelic range of 196-220 in *D.cespitosa* whereas its size in barley is noted as 129 bp, a difference of 63 bp. Previous studies have shown that mutations at microsatellite loci are not solely restricted to the hypervariable region and can occur in the flanking regions at nonnegligeable rate, both of which can contribute to variations or lack thereof in allele size (Chapuis and Estoup, 2006). As such, inferring complete sequence homology based solely on the presence of amplification product is premature.

4.2.1 Genetic diversity

Mine tailings are typically a difficult medium for plant establishment and growth as these sites often contain elevated levels of metals, low nutrients and organic matter as well as being subjected to wind and water erosion. Some plant species and/or adapted populations have successfully colonized these toxic environments, however such inhospitable conditions often leave these areas with only scattered patches of vegetation (Mining in the Yukon). *Deschampsia cespitosa* has shown a remarkable ability to colonize and dominate such plots of land with great success having naturally colonized over 1000s of hectares of barren lands around Sudbury, following the constructions of the Super Stack in 1972. As a direct result of the mining activity in both regions, Cobalt and Sudbury present extremely hostile environments that are believed to have imposed strong selection pressures on colonizers resulting in reductions in genetic diversity. Detailed analysis of our nine populations with microsatellite markers reveals the observed mean heterozygosity (Ho) and the expected mean heterozygosity (He) ranged from 0.413 and 0.48 in the Mississagi Lighthouse population to 0.645 and 0.76 in the Cobalt-3 region. In addition, genetic diversity measures based on Nei's and Shannon's index demonstrated a similar patter with values ranging from 0.39 and 0.84 in Mississagi Lighthouse to 0.61 and 1.21 in Cobalt-3. As far as soil analysis, the Cobalt-4 site was shown to be the most contaminated of all the sites with significantly higher levels of arsenic, lead, zinc, cadmium and cobalt whereas the Mississagi Lighthouse and Little Current consistently grouped as the sites with the significantly least amount of metals. Cobalt-3 typically placed in the middle of the spectrum among contaminated sites. Therefore, based on our microsatellite data, the level of genetic diversity at the population level does not decrease in terms of increased metal contamination. These findings are in conjunction with the results reported by Bush and Barrett (1993) on isozyme diversity that indicate that mine populations were no less polymorphic than uncontaminated populations. The retention of such elevated levels of genetic diversity within these mining populations can be attributed to number of selective, reproductive and demographic factors. As described by Bourret et al. (2007) if tolerance to the adverse environmental condition increases as a function of individual heterozygosity and/or if the contaminant is a mutagen, genetic variation within the affected population will remain elevated and may increase. Also, this species is a wind-pollinated outbreeder and, as a result, founders from such outbreeding populations are likely to heterozygous at many loci. In turn, this enhances the gene pool of small, founding populations by increasing the probability that at least the common alleles in the source population are represented in the new population (Bush and Barrett, 1993). Examination of the Hardy-Weinberg equilibrium across the nine populations revealed only two deviating populations, Walden and Cobalt-3 which were identified as having a significant heterozygote excess with values of 0.0021 and 0.0060 ($p<0.05$), respectively. Further analysis revealed inbreeding coefficients (F_{IS}) ranging from -0.346 in Walden to -0.19 in Cobalt-3 to -0.066 in Cobalt-4. These presences of these negative values across all nine populations imply a substantial amount of outbreeding, which as discussed earlier is in agreement with the reproductive pattern of the species. These findings also explain the occurrence of such highly heterozygote saturated populations because as stated above, outbreeders are more likely to be heterozygous at many loci (Bush and Barrett, 1993). Finally, the degree of differentiation among population (F_{ST}) was measured and was found to vary between moderate genetic differentiations with a value of 0.096 at locus HVM5 to very high with a value of 0.298 at locus HVM3. The mean degree of population differentiation was 0.194 in the *Deschampsia* population analyzed, indicating that 19.4% of the total genetic diversity is attributed to differences among populations.

4.2.2 Gene flow

Gene flow was examined to give an estimate of the average migration between all the populations studied per generation. The mean level of gene flow (N_M) in *Deschampsia cespitosa* based on our microsatellite analysis was 1.04 which is interpreted as the absolute number of individuals exchanged between populations. The level of genetic differentiation of 0.19 is regarded as high genetic differentiation between populations. It is also inversely proportional to Nm because as gene flow between populations increase, the genetic differentiation between these populations would decrease as a direct result. The low level of gene flow can be explained by the geographic distance between the nine populations, as the two closest sites are 2.1km away from each other and the two most distant sites are 319 km, despite the wind-pollinating reproductive strategy of the species.

4.3 Phylogenetic relationship

Studies by Bush and Barrett (1993) support the hypothesis that the metal-tolerant populations of *D. cespitosa* evolved at least twice in recent evolutionary history based on isozyme analysis. Secondary to their work, RAPD analysis (Nkongolo et al., 2001) of these same populations reveal a relatively small genetic distance between the four Sudbury populations which suggest that they are the results of a single colonization event. The Cavalli-Sforza and Edward's (1967) chord distance, D_C, was used to estimate the genetic distance among our nine *D. cespitosa* populations. This particular algorithm is relatively unaffected by the presence of null alleles with low to moderate frequency and it relies on allele frequencies in order to determine the geometric placement of populations in a multidimensional sphere, rather than a mutational model (Chapuis and Estoup, 2007; Khasa et al., 2006). The distance matrix based on our microsatellite data revealed that the Walden and Coniston populations were the most genetically closely related populations even though their geographic locations were not the closest (18.5km), whereas the populations from Coniston and Cobalt-3 exhibited the greatest genetic distance despite the fact that these two populations were not the furthest geographically (133.7km). The four *Deschampsia* populations from the Sudbury region clustered together along with the Mississagi Lighthouse population. These findings are partly in accordance with the findings of Nkongolo et al. (2000) based on RAPD analysis, which also identified the four Sudbury population as a single cluster along with Little Current as well as the ISSR data which also clustered the four Sudbury populations. As such, this lends support to the theory that these four populations are the result of a single colonization event. The dendrogram also clustered the three Cobalt populations, which is not similar to the groupings of Nkongolo et al. (2001). In fact, the analysis of microsatellite and ISSR data suggests a very close genetic relationship between Cobalt-3 and Cobalt-4, followed by Cobalt-5. This is in disagreement with the proposed hypotheses that Cobalt-3 population arose from an unspecified seed mix (Nkongolo et al. 2001). The data described in the present study tend to lend additional supports to the allozyme findings of Bush and Barrett (1993) which suggest that Cobalt and Sudbury have independent evolutionary histories. Finally, the Little Current population appears as very genetically distantly related from the Sudbury grouping. As such, it does not lend support to the Hutchinson theory which describes the possible colonization of Sudbury region through the railway (Nkongolo et al., 2001). Finally, based on the work of isozymes (Bush and Barrett, 1993), RAPDs (Nkongolo et al., 2001), ISSRs and microsatellites, the Mississagi Lighthouse and Little Current populations never cluster together, despite both being located on the island. The Mantel test did show a correlation between the genetic matrix and the geographic distance matrix, although this relationship does not seem to be based on the concentration of metal contaminants in the soil.

In conclusion, monitoring the genetic diversity of *D. cespitosa* populations has been useful in detecting trends that should alert ecologists to potential problems. The high genetic variability detected in the *Deschampsia* populations from Sudbury and Cobalt suggests that these are healthy populations with the evolutionary potential to respond favourably and adapt to changes or disturbances in the environment.

5. Acknowledgement

We express our appreciation to the Natural Sciences and Engineering Council of Canada (NSERC), Vale Limited (Sudbury) and Xstrata Limited for financial among populations.

6. References

Armstrong, J.S., A.J. Gibbs, R. Peakall and G. Weiller, 1994. The RAPDistance package. Available via http://life.anu.edu.au/molecular.software/rapd.html.

Berrang, P., D.F. Karnosky, R.A Mickler and J.P Bennett, 1986. Natural selection for ozone tolerance in *Populus tremuloides*. Can. J. For. Res. 16: 1214-1216

Blair, M.W., O. Panaud and S.R. McCouch, 1999. Inter-simple sequence repeat (ISSR) amplification for analysis of microsatellite motif frequency and fingerprinting in rice (*Oryza sativa* L.). Theor. Appl. Genet. 98: 780-792

Bourret, V., P. Couture, P.G.C. Campbell and L. Bernatchez, 2007. Evolutionary ecotoxicology of wild perch (*Perca flavescens*) populations chronically exposed to a polymetallic gradient. Aquatic Toxicology. 86: 76-90.

Buschiazzo, E., N.J. Gemmell, 2006. The rise, fall and renaissance of microsatellites in eukaryotic genomes. BioEssays. 28: 1040-1050.

Bush, E.J. and S.C.H. Barrett, 1993. Genetics of mine invasions by *Deschampsia cespitosa* (Poaceae). Can. J. Bot. 71: 1336-1348.

Cavalli-Sforza, L.L. and A. W. Edwards, 1967. Phylogenetic analysis. Models and estimation procedures. Am. J. Hum. Genet. 19: 223-257.

Chapius, M.P. and A. Estoup. 2007. Microsatellite null alleles and estimation of population differentiation. Molec. Biol. Evol. 24: 621-631.

Cox, R.M. and T.C. Hutchinson, 1979. Metal co-tolerances in the grass Deschampsia cespitosa. Nature. 279: 231-233

Deng, J., B. Liao, M. Ye, D. Deng, C. Lan and W. Shu, 2007. The effects of heavy metal pollution on the genetic diversity in zinc/cadmium hyperaccumulator *Sedum alfredii* populations. Plant Soil. 297: 83-92

Dudka, S., R. Ponce-Hernandez and T.C. Hutchinson, 1995. Current levels of total element concentrations in the surface layers of Sudbury's soils. Sci. Total Environ. 162: 161-171.

Dumaresq, C.G., 1993. The occurrence of arsenic and heavy metal contamination from natural and anthropogenic sources in the Cobalt area of Ontario. M.Sc. Thesis, Department of Earth Science. Carleton University. Ottawa, Ontario.

Excoffier, L., P.E. Smouse and J.M. Quattro, 1992. Analysis of molecular variance inferred from metric distances among DNA haplotypes: application to human mitochondrial DNA restriction sites. Genetics. 131: 479-491.

Fang, D.Q. and M.L. Roose, 1997. Identification of closely related citrus cultivars with inter-simple sequence repeat markers. Theor. Appl. Genet. 95: 408-417.

Foy C.D., R.L. Chaney and M.C. White, .1978. The physiology of metal toxicology in plants. Ann. Rev. Plant Physiol. 29: 11-566.

Gaitán-Solís, E., M.C. Duque, K.J. Edwards and J. Tohme, 2002. Microsatellite repeats in common bean (*Phaseolus vulgaris*): Isolation, characterization and cross-species amplification in *Phaseolus* ssp. Crop Sci. 42: 2128-2136

Gale M.D. and K.M. Devos, 1998. Comparative genetics in the grasses. Proc. Natl. Acad. Sci. USA. 95: 1971-1974.

Geburek, T., F. Scholz, W. Knabe, and A. Vornweg, 1987. Genetic studies by isozyme gene loci on tolerance and sensitivity in air polluted *Pinus sylvestris* field trial. Silvae Genet. 36: 49-53.

Guo, S.W. and E.A. Thompson. 1992. Performing the exact Hardy-Weinberg proportion for multiple alleles. Biometrics, 48: 361-372.

Hamrick, J.L., 1989. Isozymes and the analysis of genetic structure in plant populations. *In* Soltis D.E. and Soltis, P.S. (eds.), pp. 87-105. Isozymes in plant biology. Dioscorides Press, Portland, Oregon.

Hedrick, P.W., M.E. Ginevan and E.P. Ewing, 1976. Genetic polymorphism in heterogeneous environment. Ann. Review Ecol. Syst.7: 1-32.

Holloway, G.J., S.R. Povey and R.M. Sibly, 1990. The effect of new environment on adapted genetic architecture. Heredity. 64: 323-330.

Jones, E.S., M.P. Dupal, R. Kölliker, M.C. Drayton and J.W Forster, 2001. Development and characterization of simple sequence repeat (SSR) markers for perennial ryegrass (*Lolium perenne* L.). Theor. Appl. Genet. 102: 405-415.

Jiménez-Ambriz, G., C. Petit, I. Bourrié, S. Dubois, I. Olivieri and O. Ronce, 2007. Life history variation in the heavy metal tolerant plant *Thlaspi caerulescens* growing in a network of contaminated and noncontaminated sites in southern France: role of gene flow, selection and phenotypic plasticity. New Phytologist. 173: 199-215

Khasa, D.P., J.P. Jaramillo-Correa, B. Jaquish and J. Bousquet, 2006. Contrasting microsatellite variation between subalpine larch and western larch, two closely related species with different distribution patterns. Molecul. Ecol. 15: 3907-3918.

Li, C.D., B.G. Rossnagel, G.J. Scoles, 2000. The development of oat microsatellite markers and their use in identifying relationships among *Avena* species and oat cultivars. Theor. Appl. Genet. 101: 1259-1268.

Li, Y.C, A. Korol, T. Fahima, A. Beiles and E. Nevo, 2002. Microsatellites: genomic distribution, putative functions and mutational mechanisms: a review. Current Biol. 8: 1183-1186.

Liu, Z.-W., R.M. Biyashev and M.A. Saghai Maroof, 1996. Development of simple sequence repeat DNA markers and their integration into a barley linkage map. Theor. Appl. Genet. 93: 869-876.

Liu, K. and S.V. Muse, 2005. PowerMarker: integrated analysis environment for genetic marker data. Bioinformatics. 21: 2128-2129.

Lopes, I., D.J. Baird and R. Ribeiro, 2004. Genetic determination of tolerance to lethal and sublethal copper concentrations in field populations of *Daphnia longispina*. Arch. Environ. Contam. Toxicol. 46: 43-51.

Mantel, N., 1967. The detection of disease clustering and a generalized regression approach. Cancer Res. 27: 209-220.

MacNair, M.R., 1993. The genetics of metal tolerance in vascular plants. New Phytol. 124: 541-559.

MacRitchie, D. and G. Sun, 2004. Evaluating the potential of barley and wheat microsatellite markers or genetic analysis of Elymus trachycaulus complex species. Theor. Appl. Genet. 108: 720-724.

Mehes, M., K.K. Nkongolo, and P. Michael. 2007. Genetic variation in *Pinus strobus* and *P. monticola* populations from Canada: development of genome- specific markers. Plant Syst. Evol. 267: 47-63.

Mehes -Smith, M., Michael, P., and Nkongolo, K.K. 2010. Species-diagnostic and specific DNA sequences evenly distributed throughout pine and spruce chromosomes. Genome 53: 769-777.

Mengoni, A., C. Gonnelli, F. Galardi, R. Gabbrielli and M. Bazzicalupo, 2000. Genetic diversity and heavy metal tolerance in populations of *Silene paradoxa* L. (Caryophyllaceae): a random amplified polymorphic DNA analysis. Molecul. Ecol. 9: 1319-1324.

Muller, L.A.H., M. Lambaerts, J. Vangronsveld and J.V. Colpaert, 2004. AFLP-based assessment of the effects of environmental heaby metal pollution on the genetic structure of pioneer populations of *Suillus luteus*. New Phytol. 164: 297-303

Muller-Starck, G., 1985. Genetic differences between tolerant and sensitive beeches (*Fagus sylvatica* L.) in an environmentally stressed adult forest stand. Silvae *Genet*. 34: 241-246.

Nagaoka T. and Y. Ogihara, 1997. Applicability of inter-simple sequence repeat polymorphisms in wheat for use as DNA markers in comparison to RFLP and RAPD markers. Theor. Appl. Genet.. 94: 597-602

Nkongolo, K.K., 1999. RAPD and cytological analysis of *Picea spp*. from different provenances: genomic relationships among taxa. Hereditas, 130: 137-144.

Nkongolo, K.K., A. Deck and P. Michael, 2001. Molecular and cytological analyses of Deschampsia cespitosa populations from Northern Ontario (Canada). Genome. 44(5): 818-825.

Nkongolo, K.K., P. Michael and T. Demers, 2005. Application of ISSR, RAPD, and cytological markers to the certification of *Picea mariana, P. glauca,* and *P. engelmannii* trees, and their putative hybrids. Genome. 48: 302-311.

Nkongolo, K.K., Vaillancourt, A., Dobrzeniecka, S., Mehes, M., and Beckett, P.2008. Metal Content in Soil and Black Spruce (*Picea mariana*) Trees in the Sudbury Region (Ontario, Canada): Low Concentration of Nickel, Cadmium, and Arsenic Detected within Smelter Vicinity. Bull. Environ. Contam. Toxicol.. 80: 107-111

Qian, W., S. Ge and D.-Y. Hong, 2001. Genetic variation within and among populations of a wild rice *Oryza granulate* from China detected by RAPD and ISSR. Theor. Appl. Genet. 102: 440-449.

Raina, S.N., V. Rani, T. Kojima, Y. Ogihara, K.P. Singh, R.M and Devarumath, 2001. RAPD and ISSR fingerprints as useful genetic markers for analysis of genetic diversity, varietal identification, and phylogenetic relationships in peanut (*Arachis hypogaea*) cultivars and wild species. Genome. 44: 763-772.

Raymond, M. and F. Rousset. 1995. GENEPOP (Version 1.2): a population genetics software for exact tests and ecumenicism. J. Hered. 86: 248-249.

Röder, M.S., J. Plaschke, S.U. König, A. Börner, M.E. Sorrells, S.D. Tanksley and M.W. Ganal, 1995. Abundance, variability and chromosomal location of microsatellites in wheat. Mol. Gen. Genet. 246: 327-333.

Saha, M.C., J.D. Cooper, M.A. Rouf Mian, K. Chekhovskiy and G.D. May, 2006. Tall fescue genomic SSR markers: development and transferability across multiple grass species. Theor. Appl. Genet. 113: 1449-1458

Saitou, N. and M. Nei, 1987. The neighbor-joining method: a new method for reconstructing phylogenetic trees. Mol. Biol. Evol. 4: 406-425.

Service, P.M. and M.R. Rose, 1985. Genetic covariance among life-history components: the effect of novel environments. Evolution 39: 943-945.

Struss, D., J. Plieske, 1998. The use of microsatellite markers for the detection of genetic diversity in barley populations. Theor. Appl. Genet. 97: 308-315.

Sun, G.-L., O. Diaz, B. Salomon and R. von Bothmer, 1999. Genetic diversity in *Elymus caninus* as revealed by isozyme, RAPD and microsatellite markers. Genome. 42: 420-431

Van Straalen, N.M. and M.J.T.N. Timmermans, 2002. Genetic variation in toxicant-stressed populations: An evaluation of the "genetic erosion" hypothesis. Human Ecol. Risk Assess. 8: 983-1002.

Vekemans, X. and C. Lefèbvre, 1997. On the evolution of heavy metal tolerant populations in *Armenia maritime*: evidence from allozyme variation and reproductive barriers. J. Evolutionary Biology. 10: 175-191.

Von Frenkall-Insam, B.A.K. and T.C. Hutchinson, 1993. Occurrence of heavy metal tolerance and co-tolerance in *Deschampsia cespitosa* (L.) Beauv. From European and Canadian populations. New Phytol. 125: 555-564.

Weir, B.S. and C.C. Cockerman. 1984. Estimating F-statistics for the analysis of population structure. Evolution. 38: 1358-1370.

Williams, J.G.K., M.K. Hanafey, J.A. Rafalski and S.V. Tingey, 1993. Genetic analysis using random amplified markers. Methods Enzymol. 218: 705 -740.

Wu, L., A.D. Bradshaw and D.A. Thruman, 1975. The potential for evolution of heavy metal tolerance in plants. III. The rapid evolution of copper tolerance in *Agrostis stolonifera*. Heredity. 34: 165-187.

Yeh, F.C. and T.J.B. Boyle, 1997. Population genetic analysis of co-dominant and dominat markers and quantitative traits. Belg. J. Bot. 129: 157.

Zietkiewicz, E., A. Rafalski and D. Labuda, 1994. Genome fingerprinting by simple sequence repeat (SSR)-anchored polymerase chain reaction amplification. Genomics. 20: 176-183.

8

Phylogenetic Analysis of Mexican Pine Species Based on Three Loci from Different Genomes (Nuclear, Mitochondrial and Chloroplast)

Carlos F. Vargas-Mendoza[1], Nora B. Medina-Jaritz[2],
Claudia L. Ibarra-Sánchez[2], Edson A. Romero-Salas[2],
Raul Alcalde-Vázquez[2] and Abril Rodríguez-Banderas[1]
Escuela Nacional de Ciencias Biológicas, IPN Distrito Federal
[1]Department of Zoology
[2]Department of Botany
México

1. Introduction

1.1 Broad outline of the genus *Pinus*

Pines are the most important group of conifers. The genus *Pinus* emerged between Late Jurassic to Early Cretaceous period more than 150 million years ago (Gernandt et al., 2008). Like other conifers, pine trees are characterized by naked ovules, i.e. not protected inside an ovary. Because of this trait, pine trees do not have true fruits, bearing instead structures called cones (Arber & Parkin, 1907; Cronquist, 1968). Pines are distinguished from other conifers by needle-shaped leaves (aciculae), usually in clusters of 2 to 5, forming structures called fascicles (Farjon & Styles, 1997).

Several groups of coniferous trees are well represented in Mexico (Rzedowski, 1978). According to Martínez (1948), Farjon (1996) and Farjon & Styles (1997), Mexico was a second center of diversity of the genus *Pinus*, this country having the largest number of records of species of this genus. Both Martínez (1948) and Mirov (1967) recognized 38 species and 25 varieties or forms of pines in Mexico, but the number is now higher - over 45 species are recognized at present (Farjon & Styles, 1997; Perry et al., 1998).

The genus *Pinus* has two well-defined subgenera: on the one hand, trees with leaves having a single vascular bundle and several external resin canals, as well as fascicles with a usually deciduous sheath. These form the subgenus *Strobus*, some 15 species of which occur in Mexico, classified in two sections and three subsections (Farjon & Styles, 1997). On the other hand is the subgenus *Pinus* that is characterized by leaves with two vascular bundles and fascicles with a usually persistent sheath while still attached to the tree; in some cases the sheath persists even after the fall of the fascicle (Farjon & Styles, 1997). In Mexico, 34 species of this subgenus are recognized which are classified in two sections and six subsections according to Farjon & Styles (1997).

Pine forests in Mexico are usually heterogeneous in terms of both species and age of the trees. The latter may attain 1.2 m in diameter and 50 m in height. In natural forest (Figure 1), however, the largest trees may measure up to 1.5 m in diameter and may exceed 250 years in age (Challenger, 1998).

Fig. 1. *Pinus ayacahuite* at the National Park "Lagunas de Zempoala", Mexico.

1.2 Selected species and reproduction

During the 20th century there was a huge increase in the number of pine plantations in many countries. Species were selected with care trying to know the origin of those to be used, and genetic knowledge of seed-producing trees was required (Le Maitre, 1998). Thus began the use in pines of traits of origin imposed by countries importing seed for tree plantings (France, Japan, Australia and New Zealand). The intent was to grow higher yielding trees in the soil types and climates of countries where they are not native (India, Brazil, Australia, South Africa and the Philippines) (Busby, 1991). One such case is the use of *Pinus tecunumanii* in South Africa (Chapman et al., 1995).

Advances in biotechnology and modern genetics have furthered *in vitro* cultivation. Similarly, the use of molecular tools to reconstruct the phylogeny of pines has helped improve the efficient use of certain species (Gernandt et al., 2005). This has increased the use of Mexican species including *P. patula, P. oocarpa, P. chiapensis, P. maximinoi* and *P. devoniana* (Le Maitre, 1998).

1.3 Expansion of the use of Mexican pines

In 1963, a broad study was initiated of the options available for reforestation of vast deforested areas in Kenya and South Africa. This study recommended *P. caribaea* for use in these countries, given its tropical distribution (Styles, 1998). From that time on, the British government has financed seed collection and began collaborating with the Instituto Nacional de Investigaciones Forestales, Agrícolas y Pecurias (INIFAP) and Universidad Autónoma Chapingo in the study of Mexican pines capable of growing successfully in highly deforested areas within the British Commonwealth. Botanical and forestry studies have been focussed on the Forestry Institute (Oxford University Department of Plant Sciences) where much of the most significant information on the pines of Mexico and Central America is deposited, the book by Frajon & Styles (1997) being the fruit of this labor.

In forestry conferences held since 1963, the list of potentially usable pine tree species was enlarged, with *P. oocarpa, P. patula, P. tecunumanii, P. chiapensis, P. maximinoi* and *P. pseudostrobus* being recommended for reforestation or commercial exploitation in different countries. The use of these species has spread to other non-Commonwealth countries such as Argentina, Brazil, the Philippines, Puerto Rico and Colombia which now use Mexican pine species in their logging industry (Le Maitre, 1998).

1.4 Classification of pines

In recent years a large number of phylogenetic studies have been made of the genus *Pinus*. Thus, Strauss & Doerksen (1990) analyzed the pattern of restriction fragments, including at least one species of each of the 15 subsections proposed for the genus in the 1969 study by Little & Critchfield. Karalamangala & Nickrent (1989) studied the relationships of 14 taxa of the subgenus *Pinus* using isozyme loci. Krupkin et al. (1996) conducted a phylogenetic analysis of species of the subgenus *Pinus*, based on RFLP of chloroplast DNA. Later, Liston et al. (1999) constructed a phylogeny of the genus using sequences of internal transcribed spacer (ITS) regions of nuclear ribosomal DNA. Gernandt et al. (2005) made a review of the genus, proposing that sections and subsections be classified based on morphological and genetic (the genes MatK and rbcL) characters. Recently, Parks et al. (2009) performed a reconstruction of several groups of pines using the full chloroplast genome and obtaining greater resolution than with the use of other biomarkers.

The genus *Pinus* is recognized as a naturally occurring group, and the two subgenera are also widely recognized (even though their names have been changed). However, subdivisions within each subgenus remain highly controversial, and a large number of classifications have been published since Shaw's in 1914. One of our purposes in the present this study was to determine which of the latest proposals more closely approaches the evidence of markers from different genetic regions.

The most generally accepted classifications of pines, including Mexican species, are found in three studies that make use of both morphological (Farjon & Styles, 1997) and molecular - MatK and rbcL genes (Gernandt et al., 2005) - information, as well as evidence from different sources including chemical, distribution, morphological and molecular data (Price et al., 1998). These classifications are shown in Table 1.

However, as Table 1 shows, despite all the studies undertaken, a number of ancestor-descendant relationships in Mexican species included within some subsections of the genus remain unclear, perhaps due to their recent origin.

With these antecedents as a background, the aim of the present study was to examine Mexican species of pines and reconstruct their phylogenetic relationships using three types of genetic material with very different characteristics. Thus, we seek to provide a summary of certain studies (including those of our study team) in order to establish the phylogenetic and evolutionary relationships of Mexican species of pines, as these are the most commonly used species in plantations outside their region of origin (Le Maitre, 1998).

	Proposed subsection		
Species	Farjon & Styles, 1997	Price et al., 1998	Gernandt et al., 2005
P. attenuata	Attenuata	Attenuatae	Australes
P. ayacahuite	Strobi	Strobi	Strobus
P. caribea	Australes	Australes	Australes
P. cembroides	Cembroides	Cembroides	Cembroides
P. coulteri	Attenuata	Ponderosae	Ponderosae
P. culminicola	Cembroides	Cembroides	Cembroides
P. devoniana	Pseudostrobi	Ponderosae	Ponderosae
P. douglasiana	Pseudostrobi	Ponderosae	Ponderosae
P. durangensis	Oocarpae	Ponderosae	Ponderosae
P. engelmannii	Ponderosa	Ponderosae	Ponderosae
P. flexilis	Strobi	Strobi	Strobus
P. greggii	Attenuata	Oocarpae	Australes
P. hartwegii	Ponderosa	Ponderosae	Ponderosae
P. herrerae	Contortae	Oocarpae	Australes
P. jeffreyi	Ponderosa	Ponderosae	Ponderosae
P. lambertiana	Strobi	Strobi	Strobus
P. lawsonii	Oocarpae	Oocarpae	Australes
P. leiophylla	Leiophyllae	Leiophyllae	Australes
P. lumholtzii	Pseudostrobi	Leiophyllae	Australes
P. maximartinezii	Parrayanae	Cembroides	Cembroides
P. maximinoi	Pseudostrobi	Ponderosae	Ponderosae
P. monophylla	Cembroides	Cembroides	Cembroides
P. montezumae	Pseudostrobi	Ponderosae	Ponderosae
P. muricata	Attenuata	Attenuatae	Australes
P. nelsonii	Nelsoniae	Cembroides	Balfourianae
P. occidentalis	Australes	Australes	Australes
P. oocarpa	Oocarpae	Oocarpae	Australes
P. patula	Oocarpae	Oocarpae	Australes
P. pinceana	Nelsoniae	Cembroides	Cembroides
P. ponderosa	Ponderosa	Ponderosae	Ponderosae
P. praetermissa	Oocarpae	Oocarpae	Australes
P. pringlei	Oocarpae	Oocarpae	Australes
P. pseudostrobus	Pseudostrobi	Ponderosae	Ponderosae
P. quadrifolia	Cembroides		Cembroides
P. radiata	Attenuata	Attenuatae	Australes
P. remota	Cembroides	Cembroides	Cembroides
P. rzedowskii	Parrayanae	Rzedowskianae	Cembroides
P. strobiformis	Strobi		
P. strobus	Strobi	Strobi	Strobus
P. teocote	Oocarpae	Oocarpae	Australes
P. tropicalis	Pinus	Pinus	Pinus

Table 1. Three infrageneric classification systems of Mexican pines. Columns show the subsection assigned by each group of authors to a given species.

2. Genomes of *Pinus* used in this study

Three noncoding loci (ITS, MatK and Nad1) were used in the analysis for the present study. Thus, we have three DNA regions that are not subject to selective restrictions. These three noncoding regions were selected from three different genomes exhibiting different forms of inheritance in pines (Petit & Vendramin 2007). The first locus used is a region formed by the internal transcribed spacers (ITS1 and ITS2) and a fragment of the gene 5.8S. The latter is a nuclear gene transmitted by both parents to offspring (Soltis et al., 2000). The second fragment is a chloroplast mutase K (MatK) that in most pine species is inherited through the father, i.e. through pollen (Soltis et al., 2000). Finally, a fragment of the intron 2 of subunit 1 of NADH dehydrogenase present in the mitochondrion was used (Nelson & Cox, 2004). This locus is inherited exclusively through the mother and provides some idea of the inheritance of female functions in pines (Mitton et al., 2000). Table 2 lists the accession numbers for each sequence in each of the species.

2.1 Procedure for obtaining the sequence of the Nad1 gene

We amplified and sequenced a fragment of the intron 2 of the Nad1 gene, which as stated previously, is a region of the pine mitochondrion. The GenBank already had sequences for MatK (Gernandt et al, 2005), ITS1 and ITS2 (Liston et al., 1999), but reconstruction had not been made using this region. The procedure used to obtain the sequence is described below.

2.2 Extraction of total DNA

Extraction was performed using the 2% CTAB (cetyl trimethyl ammonium bromide) and chloroform-isoamyl alcohol method: 500 mg of foliage material was taken from each species and macerated with liquid nitrogen to obtain a fine powder, following the Doyle & Doyle (1987) protocol.

2.3 PCR amplification

To amplify a fragment of the intron 2 of the mitochondrial gene Nad1 by polymerase chain reaction (PCR), the primers designed by Mitton et al. (2000) and Vargas et al. (2006) were used. Both oligonucleotides have an annealing temperature (Tm) of 59.5 °C. Amplification reactions were performed in a Biometra® T-Personal thermocycler under the following conditions: initial denaturation temperature 94 °C for 5 min, and 30 cycles of denaturation at 92 °C for 1 min, alignment at 59.3 °C for 1 min, extension at 72 °C for 3 min; and a final extension at 72 °C for 5 min.

The final volume of the PCR reaction mixture was 50 µL and its composition was as follows: 5 µL buffer 10X, 3 µL $MgCl_2$ 3 mM, 4 µL dNTP 400 mM, 1.5 µL nad1G15 pM, 1.5 µL 730F 15 pM, 0.2 µL albumin at 2.95 mg/mL, 0.25 µL of 1.25 U Taq polymerase, 2 µL of DNA at 25 ng/µL, and finally 32µL H_2O.

2.4 Clonation

Clonation was performed with the TOPO TA Cloning ® Five-minute cloning of Taq polymerase-amplified PCR products kit (Invitrogen™) as described below.

2.4.1 Insertion of the DNA fragment

The clonation vector was the (linear) synthetic pCR 2.1-TOPO plasmid that has genes of tolerance to kanamycin and ampicillin, which were used as selection markers. This plasmid

also has the promoter of the Lac gene (Lac P) and the Lac Zα fragment where the gene in question is inserted, since a topoisomerase is covalently connected on its 3′ extremes and an unpaired thymidine is attached on its thyroxine 274 residue via a bond with a phosphate group. The plasmid also has two origins of replication (ColE1 ori and f1 ori), which ensures vector replication.
Ligation occurs since Taq polymerase has terminal transferance activity, which adds a desoxyadenine on the 3′ extreme of the PCR product. This desoxyadenine is complementary to the thymine residue present in the vector and through base complementarity the two are aligned, promoting rupture of the topoisomerase phosphate bond and the release of energy that is coupled by the topoisomerase ligating the DNA fragments.
Reaction conditions were as follows: in a microcentrifuge tube was placed 0.5 to 4 μL of fresh PCR product, 1 μL of saline solution, sterile water to attain a volume of 5 μL, and 1 μL of TOPO® vector to obtain a final volume of 6 μL. The reaction mixture was mixed gently prior to incubation at room temperature for 5 min. It was subsequently incubated in ice until transformation.

2.4.2 Procurement of competent cells (transformation)

The strains *Escherichia coli* DH5αTM – T1 R and *Escherichia coli* Mach1 TM –T1 R were provided by the manufacturer of the kit and had attained competent state as a result of chemical treatment by the supplier. They were kept frozen until transformation.
Both strains are characterized by sensitivity to kanamycin and ampicillin. To each of the vials containing the strains was added 2 μL pCR-TOPO 2.1 vector with the insert. Contents were mixed gently and incubated in ice for 5 to 10 min, after which thermal shock was applied for 30 s at 42 °C. The vials were again placed in ice. Then 250 μL of SOC medium (supplied by the manufacturer) was added at room temperature. Vials were capped and centrifuged at 200 rpm for 1 h. Transforming cells (10-50 μL) were seeded in plates containing LB medium with 50 μg/mL of ampicillin or kanamycin, supplemented with 40 μL Xgal at 40 mg/mL and pre-incubated at 37 °C in duplicate (to ensure procurement of isolated colonies, 20 μL of SOC medium was added). The plates were incubated at 37 °C for 8 to 12 h.

2.4.3 Selection and storage of transforming cells

Transforming cells are white with a light blue tint and negative ones are a deep blue color as a result of Xgal hydrolysis.
Ten or more colonies were selected and cultured overnight in LB medium with 50 μg/mL of ampicillin or kanamycin. Subsequently, aliquots were taken and reseeded for 4 h. To optimize clonation results, several colonies were collected.
The colonies collected were inoculated in 1 to 2 mL of LB medium with 50 μg/mL of ampicillin or kanamycin and incubated to attain stationary phase. Then 850 μL of the culture was mixed with 150 μL sterile glycerol in a cryotube, storing the latter at –80 °C until further use.

2.5 Sequencing

Sequencing of the DNA fragment was carried out by the Sanger method in a Perkin Elmer ABI Prism 910 genetic sequence analizer using the PRISM TM dye terminator cycle sequencing reaction kit, at the Unidad de Biología Molecular of the Instituto de Fisiología Celular (Universidad Nacional Autónoma de México).

Species	Gene		
	Nuclear ITS	Chloroplast Mat K	Mitochondrial Nad1
P. ayacahuite	AF036981.1	AY497257.1	
P. caribea		AB063498.1	
P. cembroides	AF343994.1	AY115783.1	
P. coulteri	AF037013.1	FJ580103.1	
P. culminicola	AF343988.1	AY115776.1	
P. devoniana		DQ168622.1	JN225481
P. douglasiana	AF037012.1	FJ580063.1	JN225482
P. durangensis	AF037010.1	FJ580067.1	JN225483
P. engelmannii		FJ580070.1	AY761139.1
P. flexilis	AY430075.1	EF546711.1	
P. greggii		DQ166030.1	JN225478
P. hartwegii	AF037008.1	FJ580088.1	JN225484
P. herrerae		AB080943.1	JN225479
P. jeffreyi	U88040.1	FJ580107.1	JN225485
P. lambertiana	AF036990.1	DQ168638.1	
P. lawsonii		AB097784.1	JN225480
P. leiophylla	AF037017.1	AB081085.1	AY761136.1
P. lumholtzii	AF037026.1	AY497278.1	
P. maximartinezii	AF036994.2	DQ168631.1	JN225470
P. maximinoi		AB161010.1	JN225486
P. monophylla	AF343986.1	DQ168632.1	
P. montezumae	AF037009.1	FJ580090.1	AY761137.1
P. muricata		FJ580111.1	
P. nelsonii	AF343999.1	DQ168633.1	
P. occidentalis		AY497281.1	
P. oocarpa		DQ353710.1	JN225472
P. patula	AF037019.1	AY497284.1	JN225473
P. pinceana	AF343996.1		
P. ponderosa	AF037011.1	FJ580108.1	AF231325.1
P. praetermissa		DQ353711.1	JN225475
P. pringleii		AY497283.1	JN225474
P. pseudostrobus		FJ580102.1	JN225487
P. quadrifolia	AF343991.1	AY115771.1	
P. radiata		AB080934.1	
P. remota	AF343989.1	AY313936.1	
P. rzedowskii	AF036996.2	AY115791.1	
P. strobiformis		EF546726.1	AB455848.1
P. strobus	AY430064.1	AY497255.1	AB455849.1
P. teocote	AF037018.1	AY497285.1	AY761138.1
P. tropicalis	AF037005.1	AB080920.1	

Table 2. GenBank sequences of each of the regions analyzed, listed by species.

3. Data analysis

3.1 Phylogenetic reconstruction

Phylogenetic trees are mathematical structures or models showing the evolutionary history of the group under study. Such trees are formed of nodes interconnected by branches. Terminal nodes are operational taxonomic units (OTUs) and represent the genomes, species or sequences being studied. Internal nodes represent hypothetical ancestors, while the ancestor of all the nodes is the root of the tree. The number of branches adjacent to a given internal node determines the degree of resolution of the node. Thus, if there are more than three branches, the node is said to be unresolved (polytomy) and may represent synchronous divergence (simultaneous evolution of more than two descendant nodes) or poor certainty of the evolutionary relationships within this group of nodes (Hall, 2008).

There are different methods for constructing phylogenetic trees, based on the known molecular or morphological information for each OTU, including distance (neighbor-joining), parsimony, maximum likelihood, and Bayesian methods (Hall, 2008). Nei & Kumar (2000) mention that selection of a particular method to reconstruct phylogeny often depends on the personal preferences of researchers or on their knowledge of a given area. Thus, researchers who are used to working with discrete morphological characters often use parsimony methods (Hall, 2008) while molecular biologists and geneticists prefer the use of analytical techniques such as maximum likelihood or Bayesian methods.

However, different authors have shown that there is no "the best method" for any one case in any one group of organisms (Nei & Kumar, 2000; Felsenstein, 2004; Hall, 2008), and the decision to use a particular method may therefore depend on software efficiency, speed of analysis, the data available, or the biological characteristics of the group being analyzed. Thus, in the end, this decision may be made based on the time available for calculation, the number of genes to be analyzed or the certainty of being able to recover the best evolutionary history of a group.

3.2 Phylogenetic reconstruction of pines using DNA sequences

To reconstruct the phylogeny of pines using the three loci from different genomes in the group of pines, it was decided to use the following strategy: first, each gene was independently analyzed and then, in order to recover the best phylogenetic signal, the three regions were combined. As these are noncoding regions, codons were not used as a weighting criterion (Nei & Kumar, 2000).

Sequences were aligned with Clustal X software taking the following parameters into account: gap opening, 15; gap extension, 6.66; and DNA weight matrix: Clustal W 1.6 (Thompson et al., 1997). After alignment, two methods of reconstruction were used: parsimony with PAUP 4.0 beta software (Swofford, 1999) and maximum likelihood with MEGA 5 software (Tamura et al., 2011).

Parsimony analysis was carried out with a branch and bound search with unordered data of equal weight and ACCTRAN branch length optimization was done. Gaps were treated as missing data. The maximum number of retained trees was 10,000. The branch support was tested with tree-bisection-reconnection (TBR) algorithm with a 1,000 resampling bootstrap using the 4.0 version of PAUP (Swoford, 1999).

Prior to maximum likelihood analysis, the MODELTEST v3.6 program (Posada and Crandall, 1998) was run in order to calculate the best substitute model for each locus and for combined analysis. The model with the highest Akaike information criterion (AIC) value was used in each case. Once the model was selected, reconstruction was performed with a heuristic maximum likelihood method using the nearest neighbor interchange (NNI) algorithm with a bootstrap of 1000 replicates in MEGA v5 software (Tamura et al., 2011).

4. Phylogenetic analysis results

The number of species analyzed per genetic region was not homogeneous since some species could not be sequenced in all cases. Nevertheless, the number of species and of variant and informative sites in all loci was adequate for parsimony and maximum likelihood analysis (Table 3).

Statistics	Gene		
	Nuclear *ITS*	Chloroplast *Mat K*	Mitochondrial *Nad1*
Number of species analyzed	23	46	36
Number of sites	3171	1718	2564
Conserved sites	632	1616	231
Variable sites	571	102	1118
Parsi Informative sites	164	66	502
Singelton sites	119	36	502
CI	0.797	0.857	0.939
RI	0.938	0.982	0.992
Number of equally parsimonious trees	174	596	514
Length	128	70	132
Best substitute model	K2+I	T92+G	K2
Log-likelihood	-182.97	-1236.41	-917.17
Ts/Tv	1.986	1.163	0.885

Table 3. General data pertaining to phylogenetic reconstruction. Two-parameter (K2) model of Kimura; three-parameter T92 model of Tamura; G = with a nonuniform rate of evolution, following instead a discrete gamma distribution; I = with a fraction of invariant sites. CI = consistency index; RI = retention index; Ts/Tv = transition/transversion ratio.

Analysis of each individual gene by each method revealed differences in formation of the groups, but the general pattern was always retained as shown in the combined analysis in Figure 1, in which reconstruction with maximum likelihood (best tree log-likelihood = -1197.22) is seen. It is evident from this phylogenetic tree that species of the subgenus *Pinus* were always separate from those of the subgenus *Strobus*. It can also be said that in the case of the subgenus *Pinus*, *P. tropicalis* is the most divergent species. If we consider some of the characteristics of this species, we see that it has a limited distribution (restricted to Caribbean islands) and is present below 700 m asl. Whether the species occurs in Mexico or not is unclear since, like *P. caribaea*, it is found only in small patches in the state of Quintana Roo.

In this same subgenus, high polytomy is seen in most species not within the *Australes* group of Gernandt et al. (2005), while the branch for this subsection is consistent and the sole difference is inclusion of *P. devoniana* (Figure 2).

As regards the subgenus *Strobus*, two well-supported groups with a bootstrap value of 99% are formed within the subsection *Strobi* of Farjon & Styles (1997) or *Strobus* of Gernandt el at. (2005); see Figure 2.

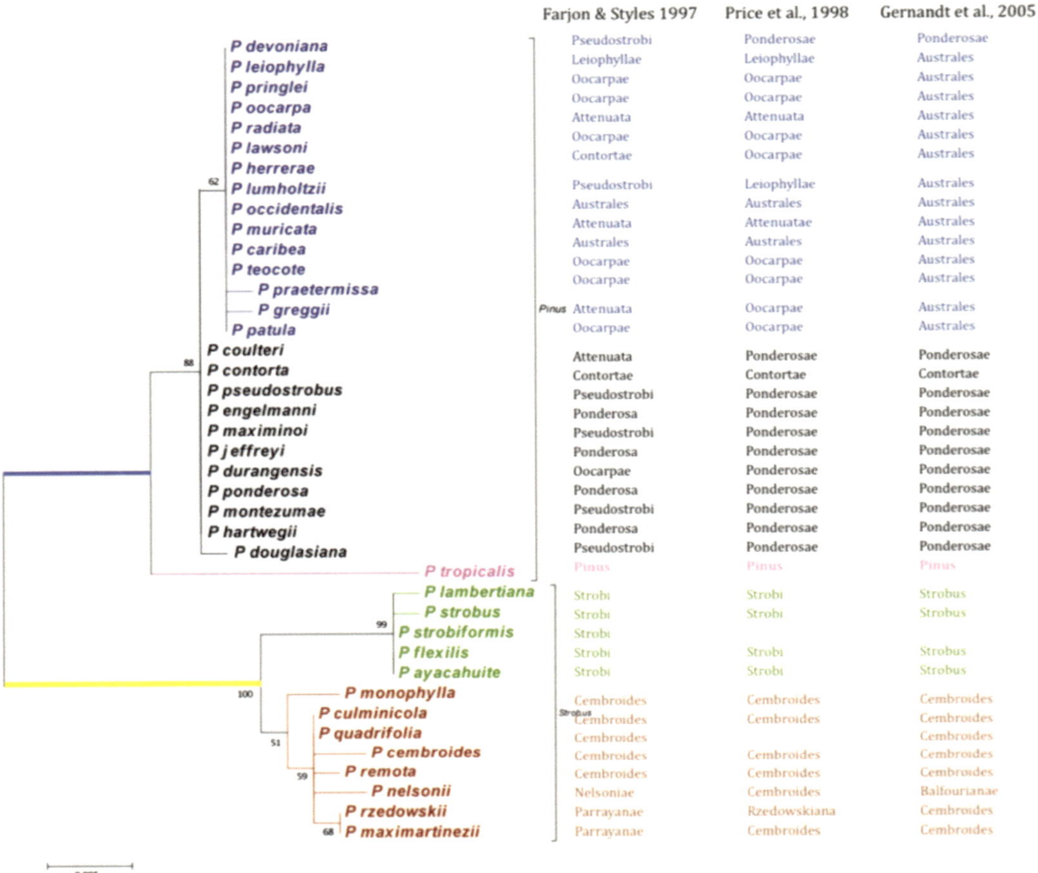

Fig. 2. Phylogenetic reconstruction by the maximum likelihood method with the T92 model (log-likelihood = -1197.22) using three loci (MatK, ITS and Nad1) from Mexican pine species. Bootstrap support is shown for each branch. Also shown are the subsections for infrageneric grouping of the pines of Mexico proposed by three groups of authors.

As regards the *Cembroides* group, *P. rzedowskii* and *P. maximartinezii* are seen to be closely related species that must be regarded as belonging to the same subsection. This branch also includes *P. nelsonii*, regarded as an independent group in all three classifications, but falling within the *Cembroides* group in the present reconstruction. The reasons for this species having been considered an independent group are to a great extent the morphological characteristics of its leaves (the three aciculae come together to form what looks like a single leaf), the shape of the scale of the seed, and a persistent peduncle in the female cone. It has a limited distribution and is therefore considered vulnerable (Farjon & Styles, 1997).

In the combined analysis of sequences, the classification proposed by Gernandt et al. (2005) may be considered to best fit the evidence derived from the three genomes of pines. Thus, the *Australes* group is consistent with a branch supported by a bootstrap value of 62%. Similarly, the *Strobus* group is well resolved since its bootstrap support is 99%. The infrageneric groups proposed by Farjon & Styles (1997) obtained the least support in this analysis, followed by those of Price et al. (1998).

5. Final considerations

5.1 Commercial use of pine trees

Trees have played a major role in the provision of goods for human communities (Le Maitre, 1998). Even today they are prevalent in agro-industry, particularly pine trees which dominate the trade in forest products (Dove, 1992).

Economic profits are so important in the lumber and related products industry that massive cultivation of pine trees has spread to more than 50 countries, most of which did not have in their native forests these species, and now being grown. In fact, Lavery & Mead (1998) say that during the 1990s more than 4 million hectares worldwide were planted with *P. radiata*, a record for this type of plant. At a cost of 35 to 45 dollars/m^3 of wood, this makes logging one of the most profitable businesses (Nelson, 2006).

In many tropical countries, the massive planting of pines has been used as a model for development. However, there are other places where pines are not considered a good crop, as these trees do not generally bear edible fruits (except for pinyon nuts) nor can they be used to feed livestock (Guggenberger et al., 1989).

Successful programs in developing countries have mixed pines with food-producing trees or grazing crops. Thus, Chile has developed management programs in which *P. radiata* is grown alongside chestnut trees (*Castanea dentata*) and pastureland for sheep (Peñaloza et al., 1985). In countries with a strong forest tradition, such as the US and Denmark, the use of mixed stands to prevent soil loss and control exploitation is very common (Adlard, 1993; Le Maitre, 1998).

The idea of sustainable and sustained development has been disseminated for a long time in societies that make use of forest resources. This vision demands not only that lumber production is not seen as an only variable but also that biodiversity in natural localities is preserved (Kumar & Nair, 2004). Thus, many areas degraded in the past by bad management are now being proposed for forest recovery with pine trees (Hof & Joyce, 1992).

However, the road to be traveled is long since the tree-farming industry continues to use large tracts of land (as in Oaxaca, Mexico), causing irreparable damage to native forests.

5.2 The importance of pine trees

Trees of the genus *Pinus* are of great ecological importance as primordial members of temperate forests. They are also economically important, being a source of wood and resins. Since earliest times humans have used pine forest products as food, medicinal remedies, building materials and fuel (Styles 1998). However, the forested areas available to the logging industry represent a small part of the planet's surface and some of them are intensively used. The United States and Canada allocate large extents of forest to the logging industry, while Mexico and Central America have fewer forest areas available (Chalenger, 1998; Perry et al., 1998).

Reduced availability of forest areas for exploitation has stimulated tree plantings for commercial purposes. In some countries such as Canada, the Russian Federation, the Philippines and Chile important natural areas have been used to grow commercially valuable species. Altered areas and even unaltered natural areas have been ravaged (eliminating flora and fauna diversity) and sown with economically profitable plants. This increases productivity while posing a risk to biodiversity and soil conservation. The same problem generated by monoculture in agriculture is now being reproduced in forest areas. Native species are being replaced or eliminated and single crops are extending into areas that were once natural ecosystems or forests (Le Maitre, 1998).

Improved market access and elimination of international borders have made many developing countries consider assimilating these methods of production, replacing their natural forests with cultivated tree stands to solve resource generation. Over 56 million hectares in 90 tropical and subtropical countries are reported to be given over to forest plantings (Le Maitre, 1998). The alternative of exploiting altered resources in natural areas in order to increase production versus allocation of large tracts to forest monoculture leaving small parcels of unaltered land is still under debate. However, in many countries this question has made governments decide for exploitation – with unpredictable consequences. Evidently, negative impacts can be prevented only if such decisions are supported with management and reforestation programs backed by adequate knowledge of the species and populations that make up the resource (Styles, 1998; Rodríguez-Banderas, 2005).
In this sense, Kimminis (1997) has described four stages in the management of forest resources that are associated with past human practice. Thus, the first stage involves solely exploitation. This is when only timber-yielding resources are extracted from a locality without any type of control or regulations. Next comes a regulatory stage in which legal and political mechanisms are created in order to establish regulations to control the rate and patterns of forest exploitation. This stage is followed by one of sustainable management, in which an ecological and evolutionary vision of forest management prevails, with timber-yielding amounts being as important as species conservation for future generations in the environment in which they have evolved. Finally, there is a social stage in which the link between local community and social interests and forest-related activities is strong, and decisions concerning forest use are made jointly and by common consent with forest owners.

5.3 Uses of Mexican pines

5.3.1 Potential management

It has been said that to have efficient management programs requires, among other things, a basic knowledge of the resource in question. This must be accompanied by the ecological characterization of species, an evolutionary knowledge (fully resolved phylogeny) of the groups to be managed in order to understand their adaptations; as well a social diagnosis of the human groups that interact with the forest (Arriaga et al., 2000).
Although the logging industry in Mexico is not as developed as in other countries, it has an important role in the national economy. Pines of the subgenus *Pinus* such as *P. patula, P. oocarpa, P. pseudostrobus* and *P. herrerae* are regarded as the most important trees for pulp and cardboard production, since their xylem forms long fibers that are useful in the manufacture of these products (Styles, 1998).
Pine trees are also important in the production of resins, particularly alpha-Pinene for cleaning products (known commercially as "Pinol"), as well as in the turpentine and cosmetics industries. In states such as Michoacán, Oaxaca, Jalisco, Chiapas and Nuevo León the use of *P. leiophylla, P. oocarpa, P. montezumae* and *P. teocote* for resin marketing is an important part of the local economy (Styles, 1998).
As regards the use of pines as a source of food, *Pinus cembroides* is the most important species, particularly in northern Mexico because of pine nut production. The latter are commercially valuable and are used to prepare diverse dishes and confectionery goods. The pink pinyon nut, obtained from Mexican species, is currently considered to have a better flavor and to be of higher quality than the white pinyon nut (Styles, 1998). *P. pinceana* and *P. nelsonii* have also been used for pinyon nut production. It is common to see the local population in Nuevo León near the end of summer and into autumn gathering pine cones in the forest and shelling the final product which is put up for sale in local markets.

5.4 Problems concerning areas of exploitation

As pines are of great importance in the forest industry, their use increases each year and has taken place indiscriminately in some localities in Mexico. Since the sawmill industry uses preferably the straightest trees with no knots, used areas of exploitation are left with remnants of poorer quality, smaller trees. This is assumed to lead to the genetic impoverishment of populations. Thus, the less healthy trees are left for forest regeneration. The result are stands with irregular crowns, winding trunks and low seed production, as has happened with *P. patula*, one of the first species to be widely used (Ledig, 1998; Styles, 1998).

Another more widespread problem in different forests in Mexico is the changeover of tracts of land with natural vegetation to agricultural areas (Perry et al., 1998; Salazar et al., 2010). This circumstance is more severe in areas where demographic pressure is strong (states such as Chiapas, the State of Mexico, Hidalgo and the Federal District).

The pressure that is exerted on certain pine species with a more limited distribution in Mexico is worth mentioning, particularly in the cases of *P. maximartinezii*, which forms a small forest of some 8 km^2 in area in Zacatecas, and *P. caribaea* in Quintana Roo, which is represented by vegetation patches covering only a few hectares (Styles, 1998). Both species are exploited locally beyond their capacity for regeneration and are likely to disappear, this circumstance being all the more serious in the case of *P. maximartinezii* as this is the only area at world level where it is found.

Our analysis indicates that most pine species with conservation problems fall within the group comprised by the subsection *Cembroides* (Farjon & Styles, 1997). To a great extent this may be due to the fact that they are pinyon pine species, the exploitation of which hampers recruitment in natural forests.

On the other hand, species belonging to the *Australes* group can potentially be used for reforestation or extensive plantings. Thus, the species *P. oocarpa, P. leiophylla* and *P. patula* are among those that have been successfully used in the lumber industry as well as for resin extraction (Styles 1998).

Fig. 3. Seedling of *Pinus* growing in a gap of a Natural Forest, Mexico.

The history of pines in Mexico goes back to ancient times, since species of the genus *Pinus* have been present in this area since the Late Cretaceous period. The presence of so many species may be due to climatic fluctuations and the vulcanism sustained by this region since that time (Miller, 1977).

Finally, we consider that studies such as this one are fundamental for the subsequent integral management of forests. Knowing which species are more closely related and which groups show a higher diversity enables such species to be considered as those of greatest potential use. Also evident are the species that are at higher risk and must be considered for adequate management and future conservation.

6. Acknowledgments

This study was financially supported by the Secretaría de Investigación y Posgrado of the Instituto Politécnico Nacional (IPN-Mexico) through project SIP 20100916, and the Consejo Nacional de Ciencia y Tecnología (CONACyT) through project 89663.

7. References

Adlard, P.G. (1993). *Historical Background.* Study No.1 Shell/WWF Plantation Review. Godalming. Surrey: Shell International Petroleum Company Limited and World Wide Foundation for Nature

Arber, E. A. N., & J Parkin. 1907. On the origin of angiosperms. *Botanical Journal of the Linnean Society* Vol. 38: 29–80 ISSN 0024-4074

Arriaga, L., J.M. Espinoza, C. Aguilar, E. Martínez, L. Gómez y E. Loa (coordinadores). (2000). Regiones terrestres prioritarias de México. Escala de trabajo 1:1 000 000. Comisión Nacional para el Conocimiento y uso de la Biodiversidad. México

Busby, J.R. (1991). BIOCLIM-a bioclimate analysis and prediction system. *Plant Protection Quaternary*, Vol.6 p.8-9, ISSN 1212-2580

Challenger, A. 1998. Utilización y conservación de los ecosistemas terrestres de México. Pasado, presente y futuro. Conabio, IBUNAM y Agrupación Sierra Madre, México

Chapman, R.A., Fairbanks D. & Louw J.H. (1995). *Bioclimatic Profiles for the Potential Afforestation of* Pinus tecunumanii *in the Eastern Transvaal.* Report FOR-DEA 815. Pretoria: Department of Water Affairs and Forestry

Cronquist, A. 1968. The evolution and classification of flowering plants. Houghton Mifflin Co., Boston, 396 p

Dove, 1992. Dove, M.R. (1992). Forester's beliefs about farmers: a priority for science research in social forestry. *Agroforestry Systems*, Vol. 17, p. 13-41, ISSN 0167-4366

Doyle, J.J. & Doyle E.S (1987). A rapid DNA isolation procedure for small quantities of fresh leaf tissue. *Phytochemical Bulletin* Vol.19, p.11-15 ISSN: 0898-3437

Farjon, A. 1996. Biodiversity of Pinus (Pinaceae) in Mexico: speciation and palaeo-endemism. *Botanical Journal of the Linnean Society*, Vol.121 p. 365-384. ISSN 0024-4074

Farjon, A & Styles, B.T. (1997). *Pinus (Pinaceae).* Flora Neotropica Monograph 75. New York Botanical Garden, New York

Felsenstein, J. (2004) *Inferring Phylogenies.* Sinauer, Sunderland MA, USA

Gernandt, D.S., López, G.G., Ortiz-García, S. & Liston A. (2005). Phylogeny and classification of *Pinus, Taxon* Vol.4, No. 1, (February 2005), pp. 29–42, ISSN 0040-0262

Gernandt, D.S., Magallon S., Lopez G.G., Flores O.Z., Willyard A. & Liston A. (2008). Use of simultaneous analysis to guide fossil-based calibrations of Pinaceae phylogeny. *International Journal of Plant Sciences* Vol.169 No.8 (Decembre, 2008) p.1086–1099 ISSN 1058-5893

Guggenberger, C., Ndulu P. & Shepherd G. (1986) After Ujmaa: farmer needs, nurseries and project sustainability in Mwanza, Kenya. Network paper-social forestry network No. 9C, Agricultural Administration Unit, Overseas Development Institute, London. Forestry Abstracts 53, No. 4441

Hall, B.G. (2008). *Phylogenetic Trees Made Easy*. Sinauer, Sunderland MA, USA

Hof, J.G. & Joyce L.A. (1992) Spatial optimization for wildlife and timber in manged ecosystems. *Forest Sciences* Vol. 82 p.489-508 ISSN 0015-749X

Karalamangala, R.R., & Nickrent, D.L. (1989). An electrophoretic study of representatives of subgenus Diploxylon of *Pinus*. *Canadian Journal Of Botany* Vol.67 p.1750-1759 ISSN 0008-4026

Kimminis, J.P. (1997). Old-growth forest: An ancient and stable sylvan equilibrium, or a relatively transitory ecosystem condition that offers people a visual and emotional feast? Balancing Act. Environmental issues in forestry. 2nd ed., UBC Press, Vancouver, Canada. 305 p

Krupkin, A. B., Liston, A. & Strauss, S. H. 1996. Phylogenetic analysis of the hard pines (*Pinus* subgenus *Pinus*, Pinaceae) from chloroplast DNA restriction site analysis. *American Journal of Botany* Vol.83 p.489–498 ISSN 0002-9122

Kumar, B. M. y P. K. R. Nair. (2004). The enigma of tropical homegardens. *Agroforestry Systems* Vol.61 p.135-152 ISSN 0167-4366

Lavery, P.B. & Mead D.J. (1998). *Pinus radiata*: a narrow endemic from North America takes on the word. Pp. 432–449 in: Richardson, D. M. (ed.), *Ecology and Biogeography of* Pinus. Cambridge Univ. Press, Cambridge

Le Maitre, D. C. 1998. Pines in cultivation: a global view. Pp. 407–431 in: Richardson, D. M. (ed.), *Ecology and Biogeography of* Pinus. Cambridge Univ. Press, Cambridge

Ledig, F.T. (1998). Genetic variation in *Pinus*. Pp. 251–280 in: Richardson, D. M. (ed.), *Ecology and Biogeography of* Pinus. Cambridge Univ. Press, Cambridge

Liston, A., W.A. Robinson, D. Piñero & E.R. Alvarez-Buylla. 1999. Phylogenetics of Pinus (Pinaceae) Based on Nuclear Ribosomal DNA Internal Transcribed Spacer Region Sequences. *Molecular Phylogenetics and Evolution* Vol.11 No.1 p. 95-109 ISSN 1055-7903

Little, E.L. & Critchfield, W.B. (1969) *Subdivisions of the Genus Pinus*. USDA Forest Service Miscellaneus Publication 1144. Washington, D.C.

Martínez, M. (1948). Los Pinos Mexicans. Ed. 2, Universidad Nacional Autónoma de México. DF. 360 p.

Miller, CN. Jr. (1977). Mesozoic conifers. *Botanical Review*. Vol.43 p.217-280. ISSN 0006-8101

Mirov, N. T. 1967. The Genus Pinus. The Ronald Press Company. New York, USA. 602 p.

Mitton, J.B., B.R. Kreiser & G.E. Rehfeldt. 2000. Primers designed to amplify a mitochondrial nad1 intron in ponderosa pine, *Pinus ponderosa*, limber pine, *P. flexilis*, and Scots pine, *P. sylvestris*. *Theoretical and Applied Genetics*, Vol.101 p. 1269-1272. ISSN 0040-5752

Nei, M. & Kumar S. (2000). *Molecular Evolution and Phylogenetics*. Oxford University Press, Oxford, New York

Nelson & Cox, 2004 13. Nelson, D. L. &. Cox M.M. 2005. Lehninger Principles of Biochemistry. Ed. W.H. Freeman, New York

Nelson, 2006 Nelson, J.D. 2003. Forest planning studio: ATLAS program reference manual version 6. URL: http://www.forestry.ubc.ca/atlas-simfor/extension/docs.html#FPS_2003

Parks, M., Cronn R. & Liston A. (2009). Increasing phylogenetic resolution at low taxonomic levels using massively parallel sequencing of chloroplast genomes. *BMC Biology,* Vol.7 p.84-101, ISSN 1471-2148

Perry, J.P., Graham A. & Richardson D.M. (1998) Pines in Mexico and Central America. Pp.137-149 in: Richardson, D. M. (ed.), *Ecology and Biogeography of* Pinus. Cambridge Univ. Press, Cambridge

Peñaloza, R., Herve M. & Sobarzo L. (1985) Applied research on multiple land use through silvopastoral system in Chile. *Agroforestry Systems,* Vol.3 p.59-77 ISSN 0167-4366

Petit, R.J. & Vendramin G.G. (2007). Plant phylogeography based on organelle genes: an introduction. Chapter 2, S. Weiss and N. Ferrand (eds.), *Phylogeography of Southern European Refugia,* p. 23 97. © 2007 Springer

Posada D, & Crandall K.A. (1998) Modeltest: testing the model of DNA substitution. *Bioinformatics* Vol.14 p.817-818

Price, R.A., A. Liston & S.H. Strauss. 1998. Phylogeny and systematics of Pinus. In Ecology and Biogeography of *Pinus* (D.M. Richardson, ed.), pp. 49-68. Cambridge University Press. Cambridge, U.K.

Rodríguez-Banderas, A. 2005. Filogenia de las especies mexicanas de *Pinus* subgénero *Pinus* subsección *Ponderosae.* Tesis de licenciatura. I.P.N. E. N. C. B. México D. F.

Rzedowski, J. 1978. *Vegetación de Mexico.* Editorial Limusa. Mexico, D.F., Mexico

Salazar C., Vargas-Mendoza C.F. & Flores J.S. (2010). Estructura y diversidad genética de *Annona squamosa* en huertos familiars mayas de la península de Yucatán. *Revista Mexicana de Biodiversidad* Vol.81 p.759-770. ISSN 1870-3453

Shaw, G. R. (1914). The Genus Pinus. Publications of the Arnold Arboretum No. 5. The Riverside Press, Cambridge, Massachussets.

Soltis, D.E., Soltis, P.S., Chase, M.W., Mort, M.E., Albach, D.C., Zanis, M., Savolainen, V., Hahn, W.H., Hoot, S.B., Fay, M.F., Axtell, M., Swensen, S.M., Nixon, K.C., Farris, J.S., 2000. Angiosperm phylogeny inferred from a combined data set of 18S rDNA, rbcL, and atpB sequences. *Botanical Journal of Linnean Society* 133, 381–461. ISSN 0024-4074

Strauss, S. H. & Doerksen, A. H. (1990). Restriction fragment analysis of pine phylogeny. *Evolution* Vol.44 p.1081–1096. ISSN 0014-3820

Styles, B.T. (1998) El género *Pinus*: su panorama en México. Pp. 385-408. In Ramamoorthy T.P., Bye R., Lot A. & Fa J. (eds.). *Diversidad Biológica de México.* Instituto de Biología UNAM, México

Swofford, D. L. (1999). PAUP. Phylogenetic Analysis Using Parsimony (and Other Methods), Version 4. Sinauer Associates, Sunderland, Massachusetts.

Tamura, K., Peterson D., Peterson N., Stecher G., Nei M. & Kumar S. (2011). MEGA5: Molecular Evolutionary Genetics Analysis (MEGA) software version 5.0. *Molecular Biology and Evolution* Vol. 24 p.1596-1599 ISSN 0737-4038

Thompson, J.D., Gibson, T.J., Plewniak F., & Higgins D.G. (1997). The Clistal X Windows interface: flexible strategies for multiple sequence alignment aided by quality analysis tools. *Nucleic Acid Research* Vol. 24: p.4876-4882 ISSN 0305-1048

Vargas, C.F., & Rodríguez-Banderas A. (2006). Evolutionary Analysis of *Pinus leiophylla*: a study using an Intron II sequence fragment of mitochondrial nad1. *Canadianl Journal of Botany.* Vol.84 p.172-177 ISSN 0008-4026